POLYMER ELECTRONICS

Contents

Preface

Conjugated polymers are a unique family of semiconductors that offer the advantages of high solubility, thin film formation in solution, and a wide range of crystallinity. They can, therefore, be used as active material for electronic as well as optoelectronic products with large areas and low cost. This book covers the topics of polymer light-emitting diodes, transistors, solar cells, and photodetectors. However, the promise of conjugated polymers is rapidly evolving and may extend far beyond these. Conjugated polymers are often combined with small molecules in order to give optimal electronic and optical properties. Many of the devices discussed in this book can also be made by solution-processed conjugated small molecules alone.

The subject of polymer semiconductor devices attracts enormous interest from researchers of diversified backgrounds, including chemistry, physics, electrical engineering, and materials science. The field is still evolving too rapidly to allow a general textbook in the common sense. The purpose of this book is to proffer a specialized textbook for students and researchers who wish to enter this exciting field. I hope that the readers will find enough background knowledge and introduction to the current research as a guide for their plans.

Finally, I wish to thank Hsiao-Wen Zan, Yu-Chiang Chao, Fang-Chung Chen, Yong-Young Noh for writing the chapters. They are all outstanding scholars of their respective subjects. I thank Hsiao-Wen Zan for helping me with the book's editing. I also wish to thank Yu-Chiang Chao, Pei-Ting Tsai, Chang-Hung Li, Hung-Cheng Lin, Po-Yi Chang, and Hsin Chiao for putting the whole material together nicely.

Hsin-Fei Meng
Winter 2012

Chapter 1

Overview of Semiconducting Conjugated Polymers

Hsin-Fei Meng
Institute of Physics, National Chiao Tung University, Hsinchu, Taiwan
meng@mail.nctu.edu.tw

1.1 Electronic Structure

1.1.1 *σ-Bond and π-Bond*

Organic molecules are based on carbon atoms by definition. The unique complexity of organic chemistry is related to two factors. First, carbon atoms are predisposed to forming chemical bonds among themselves rather than with other atoms. Like other group IV elements, carbon has intermediate electron affinity. This is in sharp contrast to, for example, alkaline atoms' low electron affinity, which tends to react with the high electron affinity of halogen atoms on the opposite side of the periodical table. Second, the carbon-carbon bond has a particularly high strength within group IV because of the carbon atom's low atomic number and small atomic size. Indeed the carbon-carbon bond energy is about 1.5 times higher than the silicon-silicon bond energy [1]. The carbon atom has six electrons in

Polymer Electronics
Edited by Hsin-Fei Meng

ISBN 978-981-4267-84-7 (Hardcover), 978-981-4364-04-1 (eBook)
www.panstanford.com

total, with two in the inner helium core and four valence electrons in the outer 2s and 2p orbitals. There are two types of common carbon bonds corresponding with the different hybridizations of 2s and 2p orbitals. In sp^3 hybridization, one spherical 2s orbital and three 2p orbitals form four superposition wavefunctions pointing to the vertices of a tetrahedron. In a diamond, each carbon atom forms four sp^3 bonds with four neighboring carbon atoms in the tetrahedral structure. Diamond has a high energy bandgap and absorbs no visible light. From a chemical point of view, both of the bonding carbon atoms donate one valence electron to the sp^3 bond. Another common type of bond is the sp^2 hybridization where the spherical 2s orbital forms superposition with $2p_x$ and $2p_y$ orbitals. All the three orbitals are symmetrical functions respective to the z-axis. In other words, they do not change sign as the coordinate z is changed to $-z$. The three hybridized orbitals point to the vertices of a triangle in x–y plane with a bond angle of 120°. In graphite, the carbon atoms have sp^2 bonds and form a honeycomb lattice, as shown in Fig. 1.1.

In such a structure, each carbon atom forms three bonds and donates three valence electrons to each of them. There are, however, four valence electrons in total in each carbon atom, and the fourth electron is in the $2p_z$ orbital, which is anti-symmetrical along the z-axis. In other words, their wavefunction has opposite signs on different sides of the x–y plane. If the carbon atoms in graphite were infinitely away from one another, the $2p_z$ orbitals would be eigenstates of the Hamiltonian system. In reality there is some wavefunction overlap among them so the eigenstates are superpositions of the $2p_z$ orbitals. Metallic energy bands result from such superpositions in graphite. In organic molecules the carbon

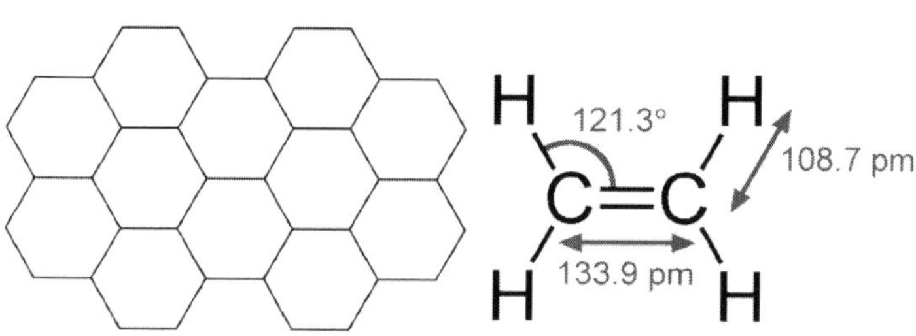

Figure 1.1 Honeycomb lattice.

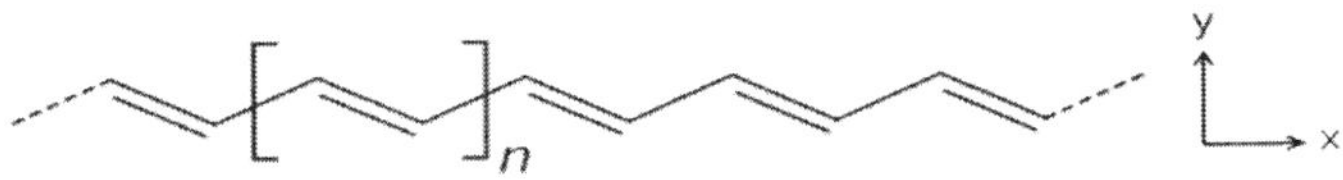

Figure 1.2 In ethylene, each carbon forms sp^2 σ-bonds with two hydrogen atoms and the other carbon atom. In addition there is a π-bond between the two carbon atoms.

atoms may form sp^2 bonds with hydrogen atoms. For example, in ethylene, each of the carbon atoms forms sp^2 bonds with two hydrogen atoms and one carbon atoms (Fig. 1.2). The carbon atom donates three electrons in all to the three sp^2 bonds, while the hydrogen atom donates its only electron to the bond. As a result, each of the sp^2 bonds contains two electrons. As before, the molecular plane is conventionally called the x–y plane. Because of the overlap, the two $2p_z$ orbitals of the two carbon atoms form linear combinations to give the Hamiltonian eigenstates. In bonding linear combination the coefficients of the two $2p_z$ orbitals have the same sign, whereas in anti-bonding orbitals the coefficients have opposite signs. Because of the absence of a nodal plane of a vanishing wavefunction between the two carbon atoms, the bonding superposition has a lower wavefunction gradient and lower energy. Each carbon atom then donates its remaining one valence electron to the sp^2 bonding orbital. In other words, in addition to the sp^2 bond there is an extra p_z bond between the two carbon atoms in ethylene, as indicated by the double bond in Fig. 1.2. The sp^2 bond has symmetrical wavefunctions along the z-axis and is usually called the σ-bond. On the other hand, the p_z bond has anti-symmetrical wavefunctions and is usually called the π-bond. The anti-bonding $2p_z$ orbital is usually indicated as π^* molecular orbital.

1.1.2 *Conjugation and π-Energy Band*

Polymers are long carbon chains with a repeat unit. They have a periodical structure in one dimension. Conjugated polymers are the ones where the carbon atoms are in sp^2 hybridization. For example, in polyacetylene each carbon atom forms two sp^2 bonds with the neighboring carbon atoms and one sp^2 bond with a hydrogen atom (Fig. 1.3). The band structure resulting from such σ-bonds has a

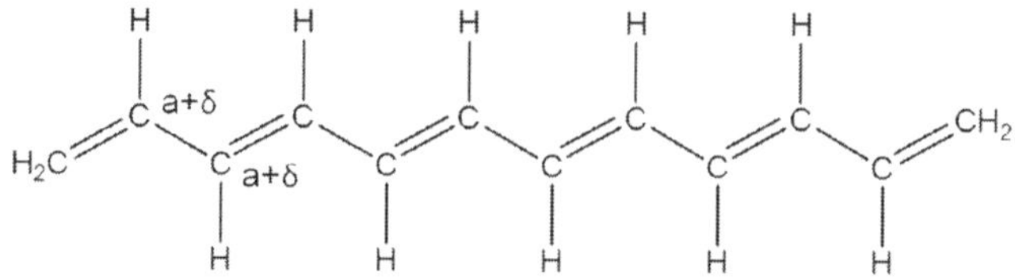

Figure 1.3 Polyacetylene has a repeat unit of C_2H_2 [2].

high energy gap and can hardly be excited. The bandgap is roughly the energy splitting between the bonding and anti-bonding σ-bonds. Because of the large overlap between the sp^2 orbitals, such splitting is large. The planar σ-bonds just provide a backbone for the polymer structure, but are not electronically active in most circumstances. In contrast to the σ-bond, in the π-bond the wavefunction overlap between the neighboring $2p_z$ orbitals is much smaller because the electron density basically extends in a normal direction to the x–y plane. The energy splitting between the bonding and anti-bonding orbitals is, therefore, much smaller and lies in the range of 1–3 eV. As in the case of graphite, Hamiltonian eigenstates are superpositions of the $2p_z$ orbitals because of the overlap. If the distance between the carbon atoms is all the same, simple tight-binding calculation gives energy bands and Fermi level, as shown in Fig. 1.4. There is a non-zero density of states at the Fermi level. In other words, the highest occupied molecular orbital (HOMO) right below the Fermi level and the lowest unoccupied molecular orbital (LUMO) right above the Fermi level are infinitely close to each other. In reality, however, the bond lengths of polyacetylene are not equal. Instead there is an alternation of short bond and long bond, as shown in Fig. 1.4. The deviation of the bond length from the natural length of σ-bonds causes an increase in the bond energy. On the other hand, such alternation opens a gap in the π-band density of states, as shown in Fig. 1.4. In fact the lattice constant is changed from a to $2a$ due to the alternation, and the first Brillouin zone boundary shrinks from π/a to $\pi/2a$. Periodical lattice potential in general opens a gap at the first Brillouin zone boundary [4]. The Fermi level now lies within the gap. Because only the levels below the Fermi level are occupied, the opening of the gap lowers the total electronic energy of the π-band. The actual degree of bond alternation is

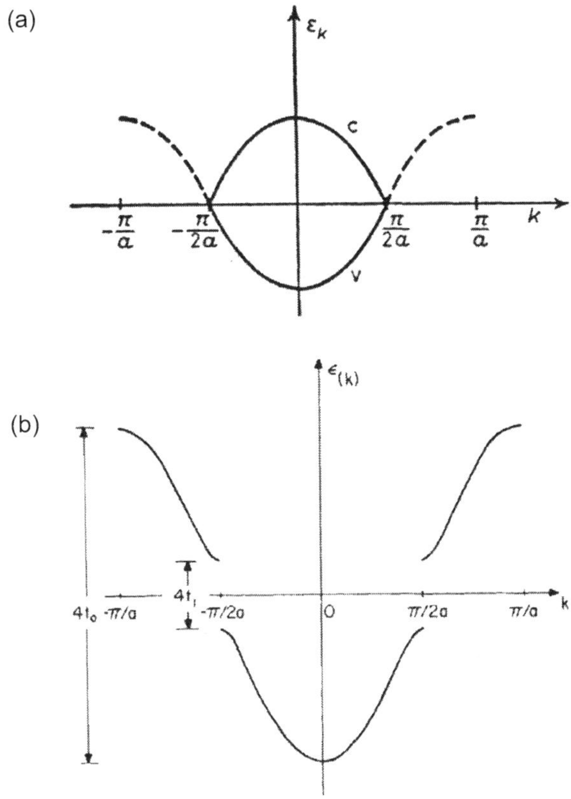

Figure 1.4 (a) If all the bonds of polyacetylene were equal in length there would be no gap [3]. (b) In reality, bond alternation opens a gap which makes polyacetylene a semiconductor rather than a metal [2].

determined by the minimum of total energy, which is the summation of increasing σ-bond energy and decreasing π-band energy. Such transformation from an equal bond length configuration to the alternating configuration is called dimerization, as the two carbons connected by a short bond form a dimmer. In chemistry convention the short bond is indicated as a double bond, and the long bond by a single bond, as shown in Fig. 1.3. One, however, has to bear in mind that the electrons are not localized in any particular bond but have wavefunctions as the linear combination of the $2p_z$ orbitals in the entire polymer chain. In the valence band wavefunctions the

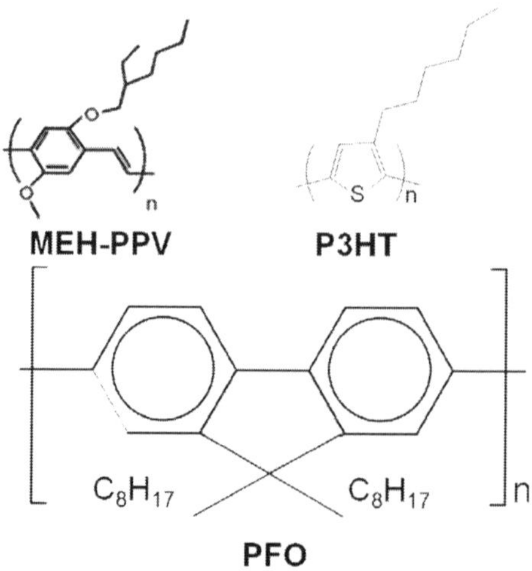

Figure 1.5 Chemical structures of MEH-PPV, PFO, and P3HT. The IP and EA values are also given in eV.

coefficients of two adjacent carbon atoms are more in phase in the linear combination. In the conduction band the coefficients are more out of phase. The valence band is therefore indicated as π-band while the conduction band is indicated as π^*-band. Similar to the bonding and anti-bonding orbitals of polyethylene, the energy gap is indicated as π–π^* gap. The concept of π-bands of polyacetylene can be easily generalized to other conjugated polymers with a more complex repeat unit, for example, poly(*p*-phenylene vinylene) (PPV), poly(9,9-dioctylfluorene) (PFO), and poly(3-hexylthiophene (P3HT), whose chemical structures are shown in Fig. 1.5.

The synthesis of conjugated polymers is covered by extensive review [5] and is not discussed below. The energy bands from simple tight-binding calculation are shown in Fig. 1.6 for PPV, PF, and PT. The number of bands is equal to the number of carbon atoms in the repeat unit as each carbon atom contributes one $2p_z$ orbital in the superposition. The bandgaps of these polymers are in the range of 1.5–3 eV. Note that there is a bond alternation in the minimum total energy state as in the case of polyacetylene. The conduction minimum is the LUMO level, and the valence maximum is the HOMO level. The ionization potential (IP) is the difference between the

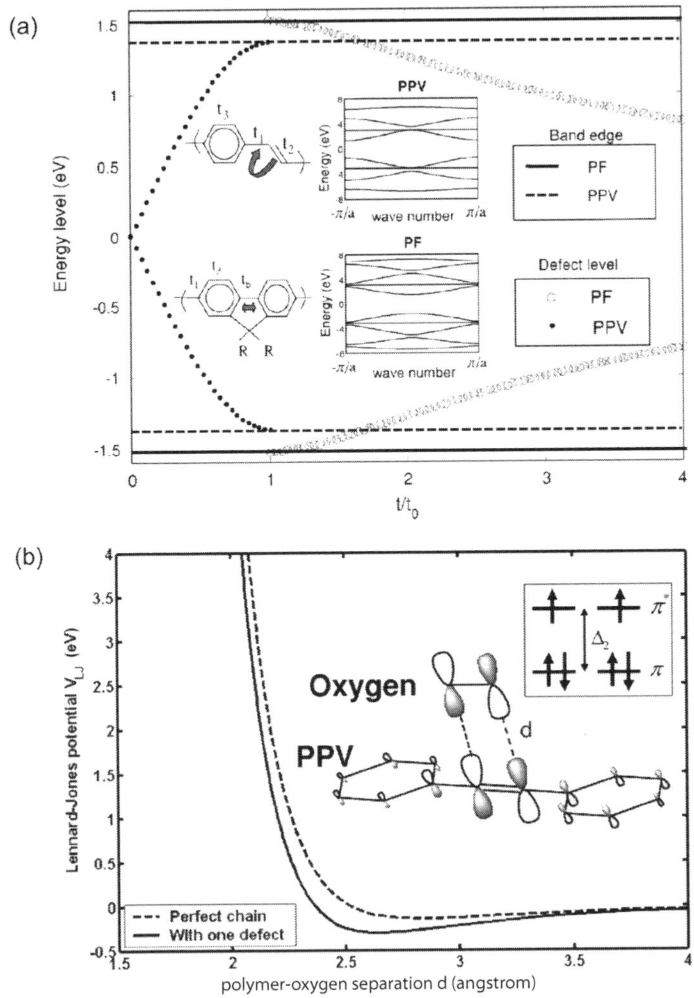

Figure 1.6 (a) The energy bands of PPV and PF [9]. (b) The energy bands of P3HT [10].

HOMO level and the vacuum level, while the electron affinity is the difference between the LUMO level and the vacuum level. The bandgap is, therefore, the difference between IP and electron affinity (EA) values. The IP and EA values for poly[2-methoxy,5-thyl-(2′-hexyloxy)-*para*-phenylenevinylene] (MEH-PPV), PFO, and P3HT are given in Fig. 1.5. The energy bands in Fig. 1.6 are for the generic backbone structures. The actual values of IP and EA depend also on the side groups.

1.1.3 *Polaron*

By definition, semiconductors are insulators with a small bandgap. For example, Si has a bandgap of 1.1 eV and GaAs of 1.4 eV. Conjugated polymers and many conjugated small molecules are semiconductors in which electrons and holes can be created by optical excitation, electrical bias, or chemical doping, as in the case of inorganic semiconductors. There are, however, some fundamental differences between the charge carriers in inorganic semiconductors and organic semiconductors. An electron in the conduction band of, say, Si is in an extended Bloch state with a well-defined crystal momentum, *k*. The crystal lattice remains perfect with or without an electron in the conduction band. On the other hand, an electron in the conduction band of conjugated polymer often accompanies a local lattice distortion which localizes the electronic wavefunction. Such an electron with local lattice distortion is called a polaron, implying that the lattice is somehow polarized by the extra electron. As discussed above, band structure depends on bond alternation. The lattice distortion in a polaron is a deviation of the bond length alternation from its perfect periodic configuration and causes a shrink in local bandgap, as shown in Fig. 1.7. Such a shrink acts as a potential well to an electron and causes bound states [6, 7]. Indeed any potential well in one dimension has a bound state, but that's not the case for three dimensions [8]. In a simple square potential well in one dimension, there is at least one even bound state, but the odd bound state exists only when the well is deep enough. The odd bound state corresponds to a three-dimensional problem. Therefore, even if there were a local lattice distortion in three-dimensional inorganic crystals, it would not localize the electrons as easily as in one-dimensional cases. Self-localization of the electron by lattice distortion is common in organic semiconductors but rare in inorganic semiconductors. The lattice distortion usually causes two gap levels, one close to the conduction band edge and one close to

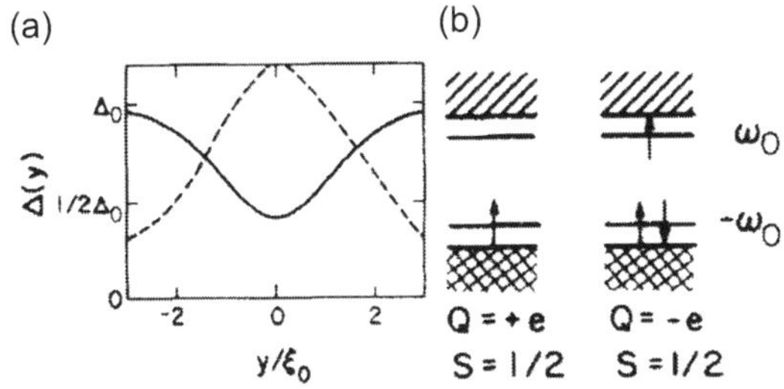

Figure 1.7 (a) Δ_0 is the bandgap of a polyacetylene chain without charge carrier. Lattice distortion around an electron causes local reduction of the bandgap. (b) Two polaron levels appear in the bandgap due to the lattice distortion [7].

the valence band edge. The lower level is squeezed out of the valence band continuum by the distortion and remains doubly occupied in case of an extra electron. The upper gap level is occupied by an unpaired electron in the electron polaron. In bipolarons the upper level is doubly occupied. Two electrons are attracted to each other by lattice distortion and merge to form one bipolaron. Similarly, in the case of hole polaron, the lower gap level is singly occupied while the upper gap level is empty as shown in Fig. 1.7. In hole bipolarons the lower gap level is empty and the polymer has two positive charges. In the presence of the electric field the polaron moves. Because the lattice distortion moves together with the localized electron, the polaron has a much higher total kinetic energy than a simple electron for given velocity. Effectively the polaron has a large effective mass due to the surrounding lattice distortion. This is one of the reasons why the carrier mobility in conjugated polymer is much smaller than inorganic crystals even if the polymer has perfect periodicity, which is of course usually not the case. In practice the carrier mobility is limited by defects and disorders, to be discussed below.

1.1.4 *Exciton*

In the ground state the valence band is filled and the conduction band empty. The low-lying excited states are excitons which are the bound state of one electron and one hole. As in the case of three-dimensional semiconductor crystals, an exciton can be viewed as an electron–hole pair with respective effective masses attracted to each other in real space by an effective Coulomb potential; only the binding energy is much larger. Indeed the one-dimensional hydrogen atom has an infinite binding energy, and some round-up of the Coulomb potential at zero relative position has to be introduced to give a finite binding energy [11]. One common theoretical approach to the excitons in conjugated polymer is the single-configuration interaction (SCI) method for the extended Hubbard model in Eq. (1.1) [12]. $t_{n,m}$ is the off-diagonal matrix element between neighboring $2p_z$ orbitals at carbon sites n and m.

$$H_0 = -\sum_{n,s} t_{n+1,n}(C^{\dagger}_{n+1,s}C_{n,s} + C^{\dagger}_{n,s}C_{n+1,s})$$

$$H = H_0 + H_{e-e}$$

$$H_{e-e} = \sum_{n} V_{n,n}\rho_n \uparrow \rho_n \downarrow + \frac{1}{2}\sum_{n}\sum_{n'(\neq n)}\sum_{s,s} V_{n,n'}\rho_{n,s}\rho_{n',s'} \quad (1.1)$$

U is the on-site Coulomb repulsion energy and $V_{n,m}$ is the long-range Coulomb repulsion energy. $C_{n,s}$ and $C^{+}_{n,s}$ are the annihilation and creation operators for an electron in carbon site n with spin quantum number $s = \pm 1/2$. The Hubbard Hamiltonian is in the second-quantization notations used for quantum many-particle system. Readers unfamiliar with second quantization need to keep in mind that conjugated polymer is a quantum many-body system with many π-electrons interacting with one another. Neglecting the interaction terms related to $V_{n,m}$ and U first, the single-particle energy levels can be obtained and they form a valence band and a conduction band, usually with a direct bandgap. The ground state has a filled valence band and empty conduction band. The lowest excited states are the ones with an electron in the valence band promoted to the conduction band. Because of the crystal momentum conservation, we need to consider only the states with a vertical promotion, shown in Fig. 1.8, where the valence and conduction band states share the same crystal momentum, k. An electron–hole pair state $|k >_{\text{eh}}$ contains one electron with crystal momentum k and one vacancy in valence band also with crystal momentum k. Each $|k >_{\text{eh}}$ is called a configuration. The Coulomb interaction Hamiltonian is represented by a matrix in the linear subspace

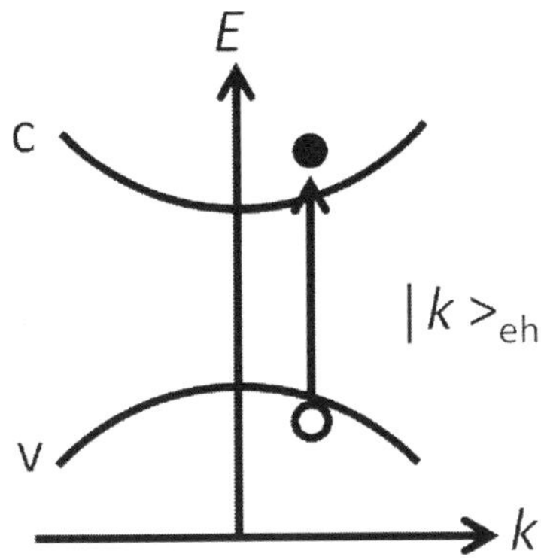

Figure 1.8 e–h pair state.

spanned by $|k>_{eh}$. Together with the free Hamiltonian involving $t_{n,m}$ only, the total Hamiltonian is solved as a matrix eigenvalue problem within the linear subspace. This problem is mathematically equivalent to the Schrödinger equation in real space with the Fourier transform of the real-space potential energy equal to the matrix element of the Coulomb interactions in the $|k>_{eh}$ subspace. Excitons are bound states in the potential well problem. As discussed above, in one dimension, there is at least one bound state no matter how weak the potential well is. The term single in SCI method means that the configurations with two electrons and two holes are not included in the subspace as they have much higher single-particle energy. In addition to model Hamiltonian, *ab initio* density functional theory is also applied to calculate the exciton states in conjugated polymers [13].

Because of the existence of a space inversion symmetry point, in most conjugated polymers all eigenstates must be parity eigenstates and are either even or odd upon space inversion. The even states are denoted by the group theory term Ag, while odd states are denoted by Bu. Since the ground state is always of Ag symmetry, only excitons with Bu symmetry have a non-zero dipole transition matrix element to the ground state. In other words, only Bu excitons can decay radiatively and emit a photon. The spin state of the electron–hole pair in SCI calculation must be specified. The electron–hole pair can be in either spin singlet state with total spin quantum number zero

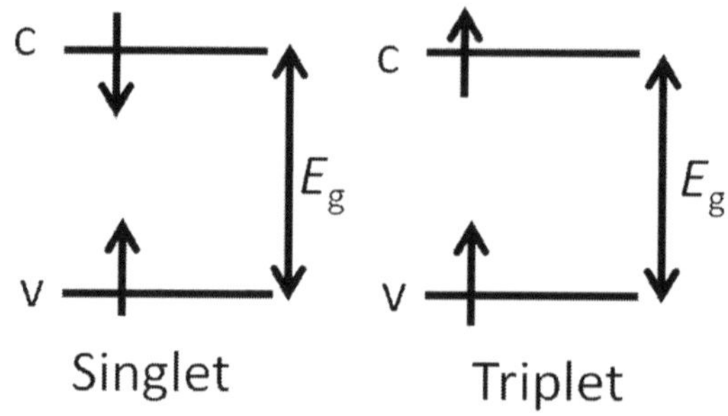

Figure 1.9 The spin state of the electron–hole pair.

or spin triplet state with total spin quantum number 1, as shown in Fig. 1.9. The electrons can be in spin-up, indicated by an upward arrow, or a spin-down state, indicated by a downward arrow.

In the singlet state the electron in the upper level has spin-down while the electron in the lower level has spin-up. On the other hand, in the triplet state both the electrons in the upper and lower level are in the spin-down states. Note spin-up and spin-down are with respect to an arbitrary quantization axis. In the singlet state the upper electron can fall down without a spin-flip, while the upper electron in the triplet state cannot fall down without a spin-flip because of the Pauli exclusion principle. There is an exchange term in the Coulomb matrix element for singlet subspace only, so the effective Coulomb attraction is weaker relative to the triplet subspace. As a result, the exciton binding energy for singlet exciton is much smaller than triplet exciton. In other words, the triplet exciton has a lower energy than the singlet in general. Their energy difference is referred to as exchange splitting. Neglecting spin–orbital interaction first, the total spin is a good quantum number of the Hamiltonian as the case of parity. Because the ground state as zero total spin and the position operator in the dipole matrix element does not involve spin, there is a non-zero dipole transition matrix element with the ground state only for singlet excitons. Triplet excitons, therefore, decay mostly non-radiatively with energy released in the form of heat instead of light.

In both SCI calculation and density functional calculation the low-lying excitons are usually ^{1}Bu, ^{1}Ag, ^{3}Bu, and ^{3}Ag states, where the numbers 1 and 3 refer to spin singlet or triplet states, respectively. The bandgap E_g can also be calculated as the onset of continuum in energy spectrum. Taking PPV as an example, SCI calculation gives E_g of 3.2 eV and ^{1}Bu energy of 2.6 eV, corresponding to singlet exciton binding energy of 0.6 eV [14]. Density functional calculation gives binding energy of 0.9 eV [13]. These numbers are dramatically larger than the binding energy in inorganic semiconductor usually in the range of 10 meV. The strong localization effect in one dimension is seen clearly. There is a quite large variation in the exciton binding energy among different calculations for PPV, but the discrepancy mostly lies in the bandgap. There is an agreement that the lowest singlet ^{1}Bu exciton is about 2.4 eV above the ground state. The above calculations are made on one isolated single chain. In solid-state film the polymer chains are arranged in three dimensions and such effect must be taken into account. Three-dimensional band structure calculation predicts the exciton binding energy of 0.3–0.4 eV [15] because the bandgap E_g is reduced by the polarization and delocalization effect. The lowest triplet exciton ^{3}Bu state for PPV is predicted to be 1.4 eV above the ground state. The exchange splitting is therefore as large as 1 eV. In contrast the exchange energy in inorganic semiconductors is negligible. The many-particle ground state and lower excited states for PPV are shown in Fig. 1.10. Above the bandgap E_g there is a continuum of free electron–hole pair states. Below the bandgap there are singlet and triplet exciton bound states with discrete levels. The exchange splitting is indicated. Only

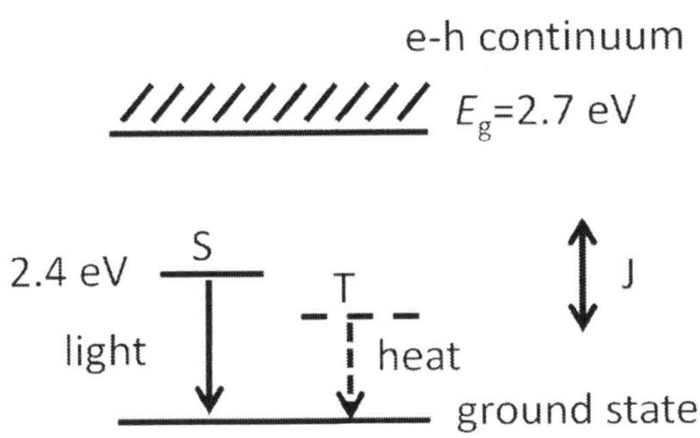

Figure 1.10 The many-particle ground state and lower excited states for PPV.

the singlet exciton decays radiatively by photon emission. Triplet exciton decays non-radiatively through multiple-phonon emission.

1.2 Optical Spectra

Upon optical excitation the conjugated polymer is excited from the ground to an exciton state. Only singlet exciton with dipole-allowed odd parity contributes to the optical absorption spectrum. Around the absorption threshold energy the excitons are derived from the valence band and conduction band. For higher photon energy the excitons are from different bands. Take PPV as an example. The valence and conduction bands are labeled D and D*, respectively [16]. Surrounding D and D* there are flat L and L* bands as shown in Fig. 1.11. There is no dispersion of the L and L* bands because the wavefunctions are localized in the phenyl rings and there is no overlap between the neighboring repeat unit. The absorption spectrum of PPV is shown in Fig. 1.11. The 2.4 eV and 4.6 eV absorption peaks are assigned to D–D* transition, the 4.8 eV absorption peak is assigned to D–L* and L–D* transition, while the 5.8 eV absorption peak is assigned to L–L* transition as shown in Fig. 1.11. The absorption peaks are quite broad due to several reasons, including the disorder in chain structure and its surrounding, coupling to phonon bath, and variation in conjugated length, which is the length of perfect conjugation. The exciton energy becomes higher in shorter conjugation segment due to quantum confinement [17]. Note that the onset of absorption at 2.4 eV agrees well with the theoretical calculations above.

Photoluminescence (PL) is the light emission following optical excitation. No matter how high the energy of initial excitation, the excitons tend to relax to the lowest states by an ultrafast process within 200 femtoseconds [18] through lattice coupling. The polymer is not photoluminescent if the ^{1}Ag exciton is lower than the ^{1}Bu exciton, as in the case of polyacetylene. For PPV and PF ^{1}Bu state is the lowest singlet exciton and strong light emission comes from its radiative decay. The exciton decay lifetime can be measured by transient photoluminescence and is about 0.3 nanoseconds. Both PPV and PF have absolute photoluminescence quantum efficiency

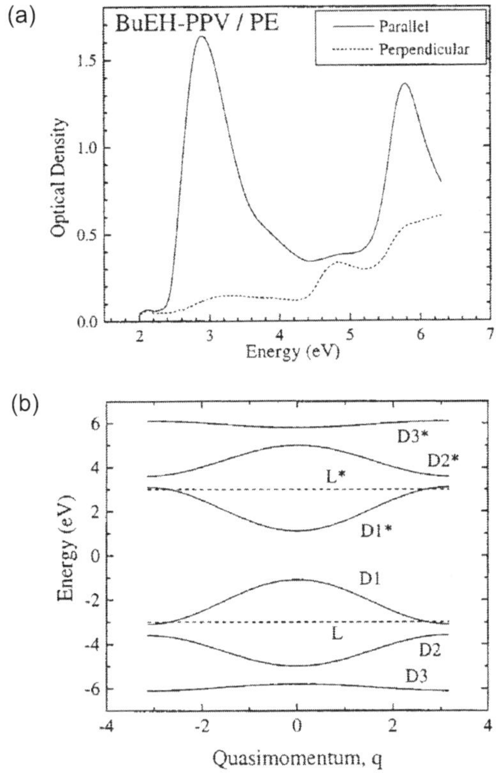

Figure 1.11 (a) The optical absorption spectrum of PPV. The onset at 2.4 eV corresponds to the singlet exciton energy. (b) The energy band labeling of PPV [16].

over 20%, i.e. 0.2 photon emitted per absorbed photon. Some of them have quantum efficiency around 40%, as shown in Table 1.1 [19].

Indeed the PL quantum efficiency of PPV and PF are among the highest for organic solids. The PL spectra of PPV are shown in Fig. 1.12. Side bands with equal energy spacing are clearly seen. For MEH-PPV the emission peak at 2.08 eV is due to the 0–0 emission of the ^{1}Bu exciton. 0–0 means that there is no phonon in either the initial state or the final state. The emission peak at 1.9 eV is due to the 0–1 transition for the same exciton, with no phonon in the

Table 1.1 The photoluminescence quantum efficiency of selected conjugated polymers

Polymer	EA (eV)	IP(eV)	PL efficiency (%)	PL peak wavelength (nm)
BP 105	3.0	5.8	30	465
Super-Yellow	3.0	5.4	22	565
PFO	3.0	5.8	40	438
MEHPPV	2.8	4.9	15	587
DPOCIOPPV	3.2	5.6	50	496
TFB	2.3	5.3	30	435

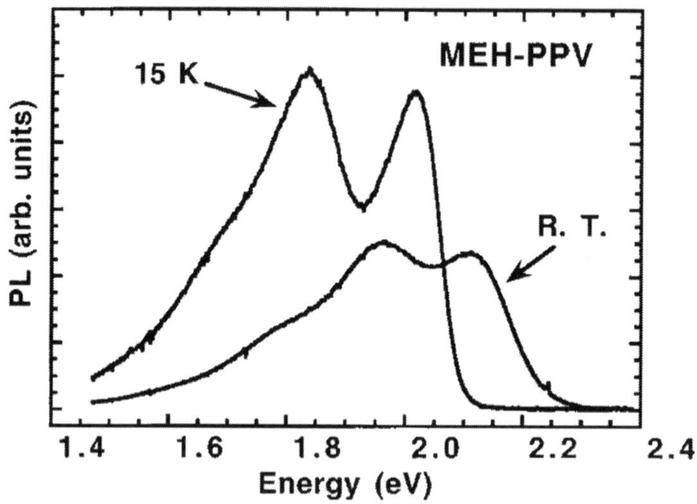

Figure 1.12 The photoluminescence spectra of PPV at room temperature and low temperature. The peak spacing is the optical phonon energy of carbon bond vibration [20].

initial state and one phonon in the final state. The photon energy is therefore smaller by the optical phonon energy of 0.18 eV.

The large phonon energy reflects the high natural vibration frequency of carbon–carbon bond due to the high bond strength and low atomic mass. Such phonon-assisted emission is possible because the shift in the minimal total energy position of the vibration coordinate in the ground state and the excited state. This is called Frank–Condon effect. Without such shift 0–1 transition would be

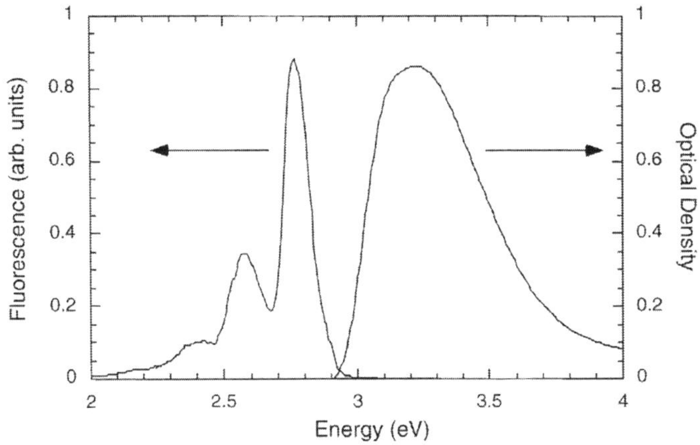

Figure 1.13 The optical absorption and emission spectra of PFO.

impossible because of the vanishing matrix element between the 0-phonon and 1-phonon states, corresponding to the ground state and first excited state of a simple harmonic oscillator. Relative strengths of the phonon side bands depend on the shift in the vibration coordinate [21]. The absorption and emission spectra of PF are shown in Fig. 1.13 [22, 23]. Similar to PPV, phonon side bands are clearly seen with a spacing of 0.16 eV. In some polymers the PL has a broad shape without the phonon side bands. The PL quantum efficiency becomes lower and the radiative decay time increases from typically 0.3 nanosecond to about 5 nanoseconds [24]. Such phenomenon is interpreted as the formation of an excimer. In solid state film the polymer chains are close to each other. Despite the disorder in large scale locally the chains tend to pack in a co-planar manner. Because of the interchain coupling the lowest excited state is no longer an exciton in one chain by a superposition of the state with one exciton in the first chain and the state with one exciton in the second chain.

Such superposition is called an excimer. The excimer energy decreases with the interchain interaction. So its presence will pull the two chains further together. Usually the excimer state has an opposite parity symmetry as the two exciton states are out of phase in superstition [25, 26]. The excimer is therefore forbidden to decay radiatively. This explains the long lifetime. The interchain nature of the excimer reduces it coupling to the intra-chain phonons, so there is no phonon side band in the emission spectrum. In CN-substituted PPV there is a particularly large red shift between PL in solution and PL in film due to excimer formation [20]. Because of the lower radiative decay efficiency excimer is to be avoided by molecular design for light emission applications.

1.3 Electrical Processes

1.3.1 *Electroluminescence*

Some conjugated polymers are highly electroluminescent. When the polymer film is sandwiched between the high work function anode and low work function cathode, electrons and holes are injected into the polymer under electric bias. The carriers move in the polymer film through drift and diffusion current, as in the case of inorganic semiconductors; only the carrier mobility is much lower. Because of the opposite charge the electron and hole capture each other by the Coulomb attraction, and eventually form an exciton. If the exciton is in the single state with odd parity it may decay radiatively and emit a photon. In fact electroluminescence has been an important reason for the great scientific interest in semiconducting conjugated polymers since 1990. After 2000, conjugated polymers drew more attention because of the photovoltaic effect.

Consider the electron injection from the cathode with low work function metal. The Fermi level of the metal is close to the conduction band edge with small barrier. At the interface electrons are injected into the organic semiconductor by thermal excitation across the energy barrier. This is called thermionic emission. The electron at the interface in the semiconductor can also fall back to the metal. This is called backflow. The backflow current density is proportional to the electron density at the interface and its velocity. The net current density through the interface is the difference between the thermionic emission current density J_{th} and the backflow current density J_{bf}. Equation (1.2) gives the conduction band density of state per unit volume.

$$
\begin{aligned}
J_{\text{th}} &= AT^2 e^{-\phi b/(kT)} \\
J_{\text{bf}}(L) &= vp(L) \\
v &= \frac{AT^2}{n_0} \\
J_p(L) &= \frac{AT^2}{n_0}[p(L) - n_0 e^{-\phi b/(kT)]}
\end{aligned}
\tag{1.2}
$$

T is the temperature, A is Richardson constant [27], and ϕ_b is the barrier height. The net current has to match the bulk current, which is the sum of drift and diffusion currents in Eq. (1.3) [28–30]. μ is the electron mobility and D is the diffusion coefficient. Einstein relation $D = \mu kT/e$ is assumed for the diffusion coefficient D .k is the Boltzmann constant, e the electron charge, x the position coordinate, E the electrical field, and $n(x)$ and $p(x)$ the electron and hole density. In general they have different mobilities. Because J_{th} is much larger than the bulk current density, the condition $J_{\text{th}} = J_{\text{bf}}$ roughly holds even under bias. This quasi-equilibrium condition determines the electron density at the boundary to be the product of the density of no state and the Boltzmann factor $\exp(-\phi_b/kT)$. This is the boundary condition to determine the current in the bulk. The smaller the Schottky barrier ϕ_b, the higher the boundary electron density, and the higher the bulk current for a given electric field. In case the injection barrier is too high for thermionic emission, Fowler–Nordheim tunneling current through the triangular potential profile may dominate the injection. Image force lowering of the barrier must be taken into account.

$$
\begin{aligned}
J_n &= e\mu \left(nE + \frac{kT}{e}\frac{\partial n}{\partial x} \right) \\
J_p &= e\mu \left(pE - \frac{kT}{e}\frac{\partial p}{\partial x} \right)
\end{aligned}
\tag{1.3}
$$

The polarons can move rapidly in a perfect linear polymer chain. In fact the effective mass of the polaron is predicted to be about ten times of the free electron mass by taking into account the lattice motion [6]. That is still a quite light mass which allows a high carrier mobility. In reality the polymer chains are disordered and the polaron motion is hindered by the interchain hopping and traps. Once a polaron reaches the end of a chain, it must tunnel through the space to the next chain, as there is no π-electron bond between the two chains. Furthermore, there are many traps due to the gap level created by various kinds of defects [31]. In PPV and PF it is often observed that the hole mobility is much higher than the electron mobility. PPV and PF contain only hydrogen and carbon atoms with relatively small electron affinity and ionization potential. In the presence of impurity with a strong electron affinity, electrons may transfer from the conjugated polymer to the impurity to form a *p*-doping. This doping leaves a hole in the polymer. Such a hole may compensate for a hole trap by filling it up with two holes [9]. The hole trap is no longer able to capture any more holes and becomes inactive. The hole mobility is therefore increased by the *p*-doping. This picture is supported by the observation that nitrogen and oxygen containing conjugated polymer or small molecules like 2-(4-biphen-4′-yl)-5-(4-*tert*-butylphenyl)-1,3,4-oxadiazole (PBD), shown in Fig. 1.14, have a relatively higher electron mobility. In fact the electron mobility is much higher than the hole mobility, unlike the case of PPV and PF. Such polymers and small molecules with a high electron affinity tend to take electrons from the impurity and become *n*-doped. The *n*-doping compensates for the electron traps by filling the defect levels with two electrons, so it can no longer capture more electrons. A common defect for the carbonyl group is

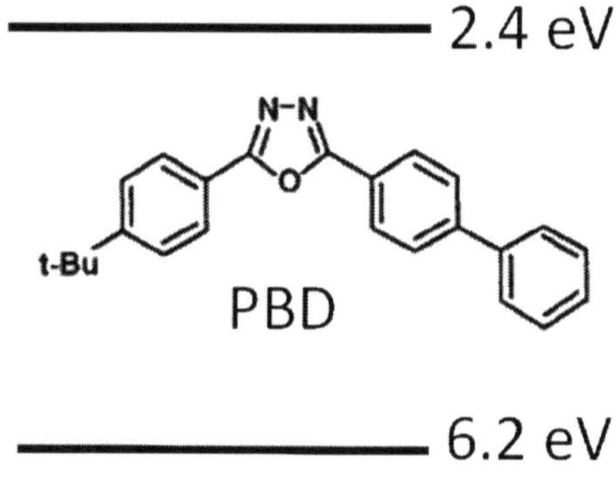

Figure 1.14 The small molecule PBD contains oxygen and nitrogen and has a higher electron affinity than most of the conjugated polymers.

shown to cause electron traps but not hole traps. This is no surprise because the extra oxygen atom tends to trap an electron due to its high electron affinity. The strong correlation between the carrier mobility ratio and the electron affinity indicates that doping plays an important role in determining the mobility.

Once an electron and a hole are so close to each other that the Coulomb attraction energy is several times higher than the thermal fluctuation energy kT, the two carriers are sure to capture each other and form an exciton. The recombination rate is expressed as $\gamma n(x)p(x)$, where γ is the recombination coefficient. In the Langevin model γ is proportional to the carrier mobility and is given by Eq. (1.4). In this model the recombination rate is basically determined by the electron flux flowing into a sphere around the hole by the drift current due to the Coulomb interaction. The flux is proportional to the mobility for a given field. The energy difference between a free electron–hole pair and an exciton is the exciton binding energy. Such energy is relaxed through cascade phonon emission during the Coulomb capture process. The continuity equation with the recombination rate as the carrier sink together with Poisson equation in Eq. (1.5) can be combined to solve the carrier density and potential distribution across the organic semiconductor film [29]. For the boundary condition the quasi-equilibrium condition above can be used for the injection side.

$$\gamma = \frac{4\pi e}{\varepsilon}(\mu_n + \mu_p) \tag{1.4}$$

$$\frac{\partial^2 \phi(x)}{\partial^2 x} = -\frac{e(p-n)}{\varepsilon} \tag{1.5a}$$

$$\begin{aligned} \frac{\partial n}{\partial t} &= \frac{1}{q}\left(\frac{\partial J_n}{\partial x}\right) - \gamma np \\ \frac{\partial p}{\partial t} &= \frac{-1}{q}\left(\frac{\partial J_n}{\partial x}\right) - \gamma np \end{aligned} \tag{1.5b}$$

In the Langevin model for recombination, one is only concerned about whether the electron and hole eventually capture each other, but not about the type of bound state that ultimately forms. As the spin of the electron and hole are arbitrary, both the spin and triplet excitons can result from the capture. Because of the triple degeneracy of the triplet exciton there are totally four possible exciton states, with one singlet higher and three triplets lower in energy, as shown in Fig. 1.10. The crucial question is then the formation ratio of the singlet and triplet excitons in electroluminescence. Since the triplet exciton does not emit light the electroluminescence efficiency depends on the ratio. If degeneracy multiplicity was the only factor that matters, the singlet formation ratio would be 25% as each of four states get an equal share. However, there is no simple reason why the ratio would be so. As discussed earlier, the capture process can be viewed as a particle moves down the Coulomb potential well via phonon cascade emission. Because the exchange interaction exists only for the singlet electron–hole pair, it experiences a weaker Coulomb attraction than the triplet electron–hole pair. Such difference in the interaction may result in a different capture cross section. Theoretical calculation based on the extended Hubbard model for two chains shows that the capture cross section for the singlet exciton is several times larger than the triplet excitons [32]. The singlet exciton formation ratio is therefore predicted to be higher than 25%. Difference in capture cross section may also result from the phonon emission process. There is a quite large energy splitting between the ^{3}Ag state and ^{3}Bu state in the triplet sector, while the splitting is much smaller in the singlet sector. This is because exchange splitting only exists between excitons with Bu symmetry, so the second lowest singlet and triplet excitons are nearly degenerate. For an electron pair in the continuum to relax into the exciton state, it has to fall across a large energy gap for the triplet sector but not for the singlet sector. The upper ^{1}Ag and ^{3}Ag levels have small binding energy, so the difference between ^{3}Ag and ^{3}Bu levels is close to the exchange splitting J. Because the exchange energy around 1 eV is much higher than the phonon energy around 0.18 eV, multiple-phonon emission with a low probability is required. This causes a phonon bottleneck for the formation of the ^{3}Bu exciton. The bottleneck may push the ^{3}Ag excitons into the singlet sector through the weak spin–orbital interaction. Formation ratio higher than 25% may result from the phonon bottleneck [33].

Experimentally triplet excitons are more difficult to detect than the singlet excitons because the former do not emit light. However, once triplet excitons are formed they cause an extra absorption due to the dipole-allowed optical transition ^{3}Ag and higher triplet levels. For PF the absorption peaks at 1.4 eV. Such infrared photo-induced absorption (PIA) can be used to study triplet excitons. Earlier works report that the singlet to triplet exciton formation ratio is over 25% based on PIA [34]. But later, it is reported that the apparent deviation from 25% is actually due to the triplet annihilation by polarons, which are also optically generated [35]. Direct measurement of the triplet exciton in electroluminescence can be done by detecting the 1.4 eV induced absorption due to electrical bias in a light-emitting diode [36, 37]. After taking into account the triplet-polaron annihilation the singlet and triplet formation rates are found to follow the same voltage dependence, suggesting a simple 25% ratio independent on the form of the interactions and the relaxation process. The singlet ratio of 25% is also established in small-molecule electroluminescence by incorporation of the heavy-metal emitter which harvests the otherwise non-emissive triplet excitons due to the strong spin–orbital interaction around the heavy-metal ion [38]. In fact it would be surprising if the singlet exciton formation ratio is different in conjugated polymer and small molecule.

Because roughly 75% of the electron–hole pairs eventually form triplet excitons in conjugated polymer electroluminescence, an iridium (Ir) complex can be added to the polymer to harvest the triplet energy and increase the light emission efficiency. The iridium complex contains an iridium metal ion at the center surrounded by a conjugated organic ligand as shown in Fig. 1.15. Iridium, with atomic number 77, is a heavy metal. Often the HOMO level is the 5d orbital of the iridium ion and the LUMO level is in the ligand. The iridium complex exciton is basically the state with an electron promoted from the iridium HOMO to the ligand LUMO, usually called a metal-ligand charge-transfer exciton. The lowest-lying exciton is in spin triplet state as in the case of conjugated polymer. The major difference between the iridium complex and purely organic semiconductor is that the former contains a heavy transition atom. Spin–orbital interaction is much stronger in the iridium complex as the interaction is proportional to the fourth power of the atomic number, as given in Eq. (1.6).

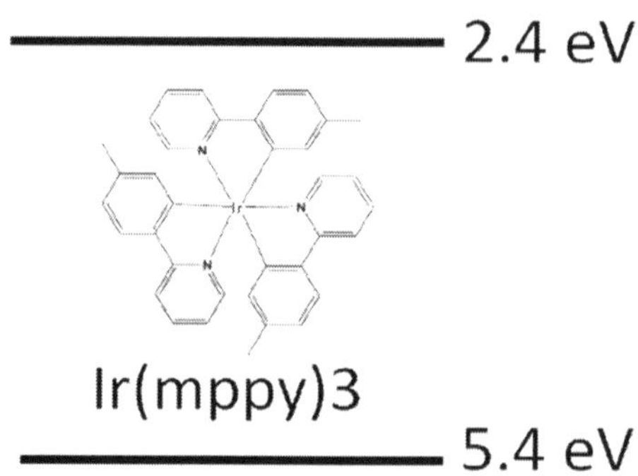

Figure 1.15 The chemical structure of green emitting iridium complex Ir(mppy)3. The HOMO and LUMO levels are 5.4 and 2.4 eV, respectively.

$$H_{\text{so}} = \frac{Ze^2}{2m^2c^2}\vec{S}\cdot\vec{L}\frac{1}{r^3}$$
$$< \frac{1}{r^3} > nt = \frac{Z^3}{a_o^3}\frac{1}{n^3 l\left(l+\frac{1}{2}\right)(l+1)} \tag{1.6}$$

S and L are the spin and orbital angular momentum operators, respectively, r is the electron position, and m is the free electron mass. The average of r^3 is for atomic orbital with principle quantum number n and angular momentum quantum number l. The spin–orbital interaction Hamiltonian H_{so} does not commute with the spin operators. So the spin–orbital interaction makes the total spin no longer a good quantum number of the total Hamiltonian. The eigenstate therefore becomes a superposition of singlet and triplet states. The lowest-lying triplet exciton of the iridium complex has in fact a small singlet component, which gives a non-zero optical transition matrix element to the singlet ground state. Unlike the case of conjugated polymer, the triplet exciton in some iridium complex decays radiatively with high quantum efficiency. Such light emission is called phosphorescence, whereas the emission from singlet exciton is called fluorescence. If the LUMO and HOMO levels of the iridium complex lie within the bandgap of the polymer, they form carrier traps in the blend of the two materials. An electron moving in the conduction band of the polymer may be trapped by the triplet emitter; then a hole is attracted toward it and trapped as well to form an triplet exciton. Phosphorescence then follows

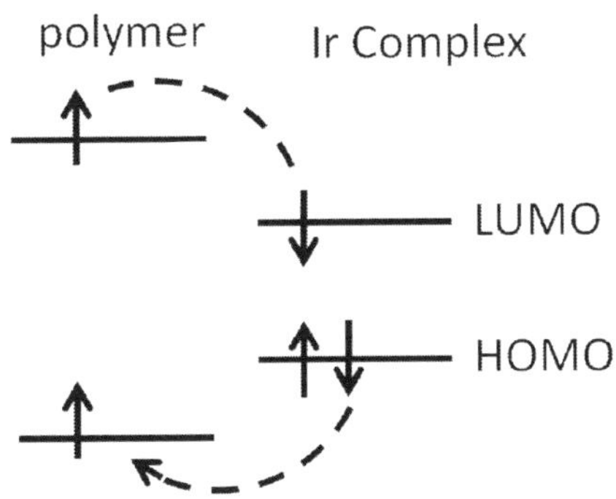

Figure 1.16 Dexter energy transfer.

as the triplet exciton decays to the ground state, with lifetime in the microsecond range. This is at least three orders of magnitude longer than the fluorescence lifetime of the polymer in the sub-nanosecond range. Triplet excitons formed in the polymer may also relax to the iridium complex triplet exciton state through a short-ranged Dexter energy transfer [38, 39] shown in Fig. 1.16. In Dexter energy transfer a spin-up electron in the polymer LUMO has a Coulomb scattering with a spin-down electron in the Ir complex HOMO. As a result the first electron is scattered to the complex LUMO while the second electron is scattered to the polymer HOMO. Some wavefunction overlap between the initial and final states for both electrons is necessary to have a quantum transition rate. So the process occurs only within a short range. Such process has only a minor contribution in the electro-phosphorescence in the polymer–complex blend. The triplet exciton level of the polymer has to be higher than the iridium complex triplet exciton. Otherwise the triplet excitons formed in the complex by carrier trapping will not decay but relax to the polymer triplet level and eventually decay non-radiatively. Prevention of such transfer by choosing polymers with a high triplet level is called triplet exciton confinement.

1.3.2 *Photoconductivity*

Charge carriers are generated upon optical excitation in solid state films of conjugated polymers. In inorganic semiconductors free electron–hole pairs are generated near the absorption energy threshold as the exciting binding energy of a few milli-electron volts is smaller than the room temperature thermal energy of 25 meV. On the contrary, the exciton binding energy of 0.3 eV for conjugated polymer is much higher than the room temperature thermal energy, so they should remain bound together. The bound exciton is in a neutral state and does not move in an electrical field, unlike the free electron and free hole. In experiments photoconductivity is, however, clearly observed even near the absorption threshold [40]. Therefore, there rises the question of how the charge carriers are generated. Phonon-assisted field electron–hole dissociation through a tunneling process has been proposed [40]. Such process is possible for hot excitons created by photons with energy much higher than the threshold as the excess energy gives high local temperature. For cool excitons without excess energy the tunneling process appears to be too weak to explain the observed photoconductivity. Defect levels, in particular the one caused by the carbonyl group shown in Fig. 1.17 [41], can act as a dissociation center for the excitons.

When the defect level within the bandgap is empty, the electron of the exciton can fall into the defect level and the hole is released and contributes to the electrical conductivity. The energy loss of the electron can be compensated by the kinetic energy of the hole. This process is called defect Auger process [42]. The external quantum efficiency of photoconductivity, defined as the number of electrons collected by the electrode per incident photon, is low for pure conjugated polymer films. It can be enhanced up to 70% by adding electron acceptors to dissociate the excitons at the polymer–acceptor interface as shown in Fig. 1.18.

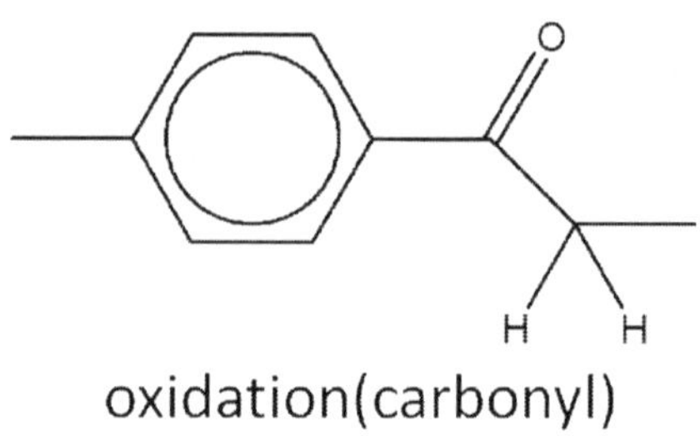

Figure 1.17 The carbonyl group due to oxidation [31].

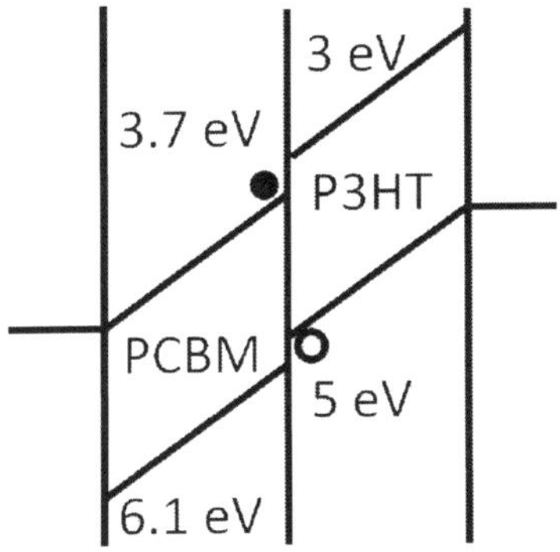

Figure 1.18 The carbonyl group due to oxidation.

1.4 Crystalline Conjugated Polymers

Most of the PPV and PF discussed above are amorphous in solid film. In fact such amorphous film morphology is required for a high solid state PL quantum efficiency. Excimers may form if two chains are packed closely, and the dipole-allowed exciton state is changed to a dipole-forbidden excimer state. The luminescence is then reduced significantly. Indeed most of the organic molecules show higher luminescence efficiency in dilute solution than in thin film, indicating the lowest excited state is somehow changed from a luminescent one in isolated molecules or polymer chains to a non-luminescent one once there is a wavefunction overlap. This phenomenon is called solid state quenching or self-quenching. In the amorphous state the side chains of the conjugated polymers protect the conjugated main chain from getting too close to each other, so the electronic states remain, as in the case of isolated chains, in dilute solution except for dielectric effect. A few semiconducting conjugated polymers have a strong tendency to self-assemble into local crystalline structures. Macroscopically the solid film is still uniform like amorphous polymers, but in the scale of 100 nm, crystalline domains form. One important example is the regioregular polymer P3HT, shown in Fig. 1.19. There are two possible carbon positions for the alkyl side chains to attach to the pentagonal thiophene ring, as indicated in Fig. 1.19. In regiorandom P3HT the

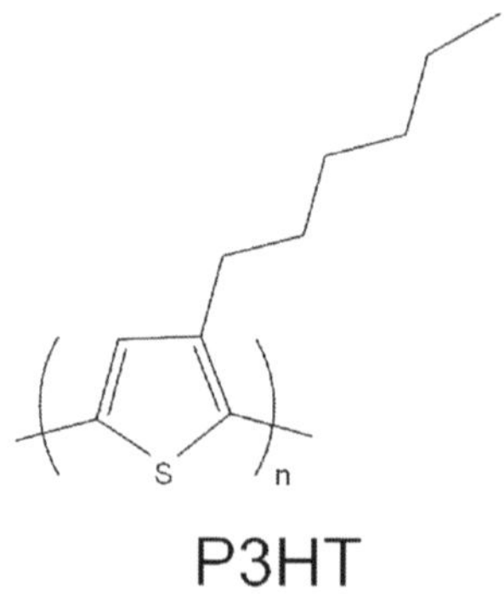

Figure 1.19 Chemical structure of regioregular P3HT.

six-carbon side chain R is attached to one of the two positions randomly along the backbone. On the other hand, in regioregular P3HT the side chain R is attached to the same position along the backbone. Regioregular P3HT is shown to have crystalline structure with the six-carbon alkyl side chains extending into each other [43], as shown in Fig. 1.20.

Below we denote regioregular P3HT by simply P3HT. The charge carrier in the P3HT film is different from the ones in other conjugated polymers discussed above. In the amorphous structure the interchain interaction is not strong, so the electronic structure of the carriers and the excitons in the film and in isolated chain are similar. However, in the crystalline P3HT a three-dimensional structure has to be considered and the charge carrier is no longer in the form of a single-chain polaron. Because the wavefunction now has a three-dimensional nature, an electron is no longer localized in a chain. Infrared photo-induced absorption has been used to detect the polaron by the optical absorption from the valence band to the lower polaron level. In P3HT the charge carriers are found to be in the form of a delocalized polaron with very small lattice relaxation energy [44]. This is consistent with the picture that self-localization is important only in one dimension, where any potential well possess at least a bound state no matter how weak it is. Apparently because of the local crystalline structure P3HT has a better hole mobility than most other conjugated polymers. The carrier transport is limited by the hopping between the crystalline

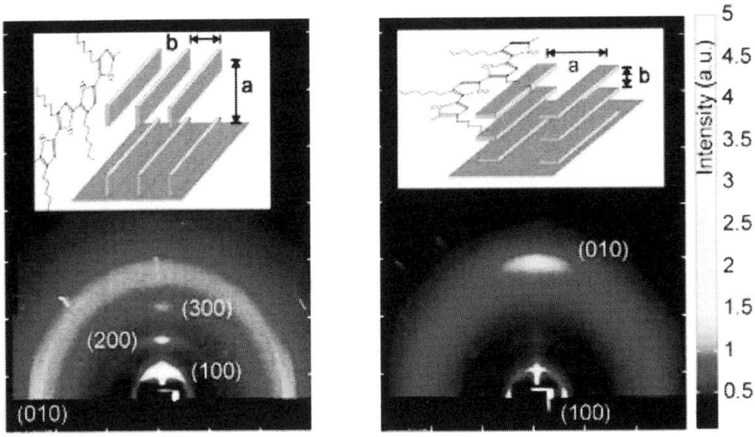

Figure 1.20 Crystalline structure of P3HT. Reprinted from *Nature*, 401, H. Sirringhaus, P. J. Browm, R. H. Friend, M. M. Nielsen, K. Bechgaard, B. M. W. Langeveld-Voss, A. J. H. Spiering, R. A. J. Janssen, E. W. Meijer, P. Herwig, D. M. de Leeuw, Two-dimensional charge transport in self-organized, high-mobility conjugated polymers, 685. Copyright 2011, with permission from Elsevier. See also Color Insert.

domains. The electron mobility is still slow, consistent with the above picture that the background *p*-doping removes the hole traps but leaves the electron traps intact. P3HT has a particularly small ionization potential of 5 eV, making it susceptible for *p*-doping from various impurities. The morphology of P3HT films depends on the substrate surface properties. Its crystallinity tends to be stronger on non-polar surfaces. For example, the polar SiO_2 surface does not favor the self-assembly of the P3HT chains because of the low affinity of the alkyl side chains to the polar OH groups on the oxide surface. Self-assembled monolayers (SAM) are often added to the oxide surface to enhance crystallinity. Hexamethyldisilazane (HMDS) and octadecyltrichlorosilane (OTS), shown in Fig. 1.21, are the most commonly used molecules for SAM. The heads of the molecules react and anchor on the oxide surface and the non-polar tail is in contact with the polymers.

Because of the low ionization potential of 5.0 eV of P3HT, it can be *p*-doped by many kinds of impurities with a high electron affinity. For example, the molecular acceptor tetrafluoro-

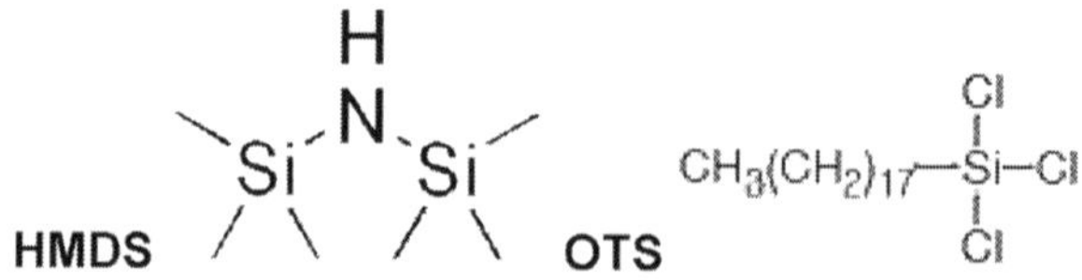

Figure 1.21 The chemical structure of the molecules commonly used for SAM to enhance ordering of polymers on polar oxide surface. The heads of the molecules react and anchor on the oxide surface and the non-polar tail is in contact with the polymers.

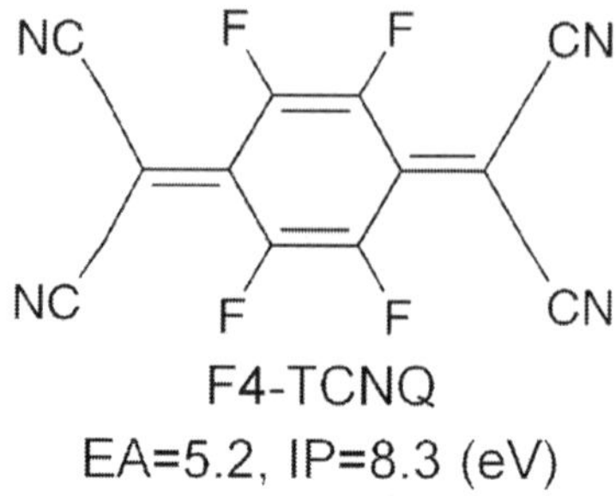

Figure 1.22 The chemical structure, EA, and IP of F4-TCNQ.

tetracyanoquinodimethane (F4-TCNQ) with a high electron affinity, shown in Fig. 1.22, is known to dope P3HT by blending together.

P3HT is also *p*-doped by the oxygen molecules in the ambient, so its electrical properties are difficult to control. The doping is studied by the capacitance of the depletion region near a metal contact [45] or by electrical conductivity [46]. Oxygen doping and de-doping are slow processes. Doping occurs in several hours, while de-doping in vacuum takes place in weeks. The doping is caused by the physical adsorption of the oxygen molecules on P3HT and not by chemical reaction, as the doping is always reversible. However, the slow de-doping process indicates that the physical adsorption strength is rather high. Heating the P3HT film to the glass temperature accelerates the de-doping process to minutes. Absorption of infrared photon around 1.8 eV below the P3HT absorption bandgap is found to accelerate the doping process to seconds. The *p*-doping of P3HT by oxygen can be understood by a wavefunction hybridization between P3HT band states and oxygen

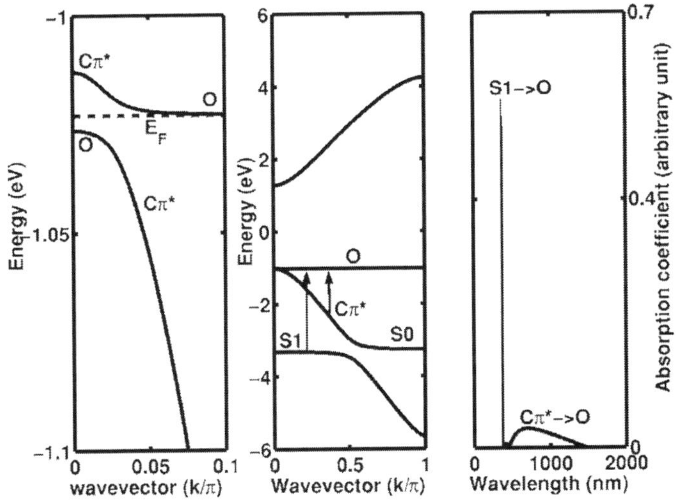

Figure 1.23 The energy bands of P3HT-O_2 combined system. Some electrons in P3HT transfer to oxygen band, so the Fermi level is below the valence band top and *p*-doping results. Extra absorption rises from oxygen adsorption as indicated by the vertical arrows [47].

molecular orbitals. The oxygen molecule has two unpaired electrons in each of the two degenerate π^* LUMO orbitals. One of the orbitals has the same symmetry upon plane reflection as the P3HT band states and becomes hybridized in the eigenstates of the P3HT-O_2 combined system. When a large amount of oxygen molecules are absorbed, there is an extra oxygen band in the total band structure as shown in Fig. 1.23 [47].

Without the coupling between the oxygen and P3HT, the oxygen band is below the P3HT Fermi level and half-filled as each oxygen molecule contributes only one electron. With the coupling the electron transfers from P3HT to the oxygen band and raises its energy through Coulomb repulsion. The Fermi level is then pushed down inside the valence band as shown in Fig. 1.23. There is therefore a non-zero hole carrier density in P3HT. The infrared absorption around of 1.8 eV [48] is due to the optical transition from the P3HT band to the nearly half-filled oxygen band. Even though there is no chemical reaction between P3HT and oxygen, the de-doping is slow because such charge-transfer state is the ground state of the combined system. This charge-transfer adsorption is much stronger than the ordinary van der Waals interaction. In addition to the doping through physical adsorption, chemical reaction between P3HT and oxygen is shown to cause an electron trap but not a hole trap [10]. This explains why the hole mobility remains high in the presence of oxygen while the electron mobility is often too low to be accurately determined. Despite superior self-assembly, good film-forming property, and simple chemical structure, the sensitivity to the ambient oxygen molecules makes P3HT not very promising as field-effect transistor for some practical purposes. Some conjugated polymers with chemical structure similar to P3HT but with higher ionization potential are developed. They retain the high carrier mobility around 0.1 cm^2/Vs but possess higher stability against oxygen doping. In these polymers there is more ring torsion and less π-orbital overlap. The bandgap becomes wider, and the valence band edge lower. The chemical structure and ionization potential of some such polymers are shown in Fig. 1.24 [49–51].

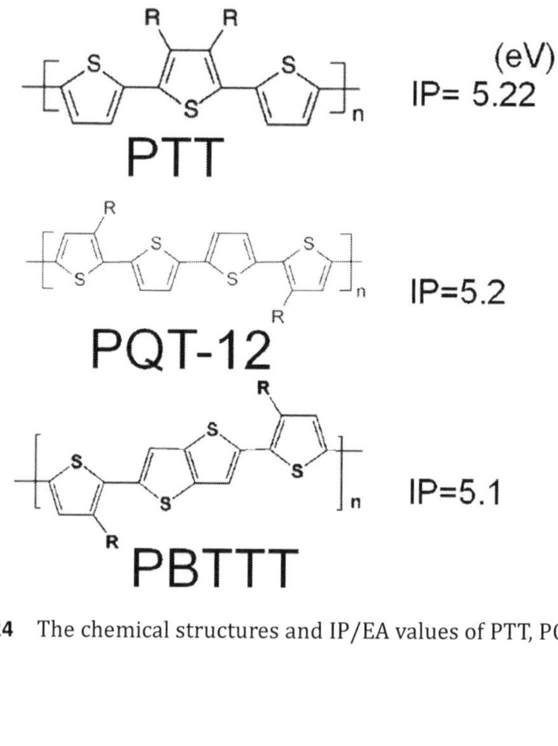

Figure 1.24 The chemical structures and IP/EA values of PTT, PQT-12, and PBTTT.

References

1. M. Silberberg, *Principles of General Chemistry*, 2nd ed., McGraw-Hill (2009).
2. W. P. Su, J. R. Schrieffer, A. J. Heeger, *Phys. Rev. Lett.* **42**, 1698 (1979).
3. A. J. Heeger, S. Kivelson, J. R. Schrieffer, W.-P. Su, *Modern Phys.* **60**, 781 (1988).
4. N. W. Ashcroft, N. D. Mermin, *Solid State Physics*, Brooks/Cole (1976).
5. A. C. Grimsdale, K. L. Chan, R. E. Martin, P. G. Jokisz, A. B. Holmes, *Chem. Rev.* **109**, 897 (2009).
6. W. P. Su, J. R. Schrieffer, A. J. Heeger, *Phys. Rev. B* **22**, 2099 (1980).
7. K. Fesser, A. R. Bishop, D. K. Campbell, *Phys. Rev. B* **27**, 4804 (1983).
8. S. Gasiorowicz, *Quantum Physics*, Wiley (1974).
9. H. F. Meng, Y. S. Chen, *Phys. Rev. B* **70**, 115208-1 (2004).
10. C. K. Lu, S. T. Pi, H. F. Meng, *Phys. Rev. B* **75**, 195206 (2007).
11. R. Loudon, *Am. J. Phys.* **27**, 649 (1959).
12. S. Abe, J. Yu, W. P. Su, *Phys. Rev. B* **45**, 8264 (1992).
13. M. Rohlfing, S. G. Louie, *Phys. Rev. Lett.* **82**, 1959 (1999).
14. M. Y. Lavrentiev, W. Barford, S. J. Martin, H. Daly, R. J. Bursill, *Phys. Rev. B* **59**, 9987 (1999).
15. P. G. da Costa, R. G. Dandrea, E. M. Conwell, *Phys. Rev. B* **47**, 1800 (1993).
16. E. K. Miller, C. Y. Yang, A. J. Heeger, *Phys. Rev. B* **62**, 6889 (2000).
17. J. Cornil, D. Beljonne, C. M. Heller, I. H. Campbell, B. K. Laurich, D. L. Smith, D. D. C. Bradley, K. Müllen, J. L. Brédas, *Chem. Phys. Lett.* **278**, 139 (1997).
18. M. Yan, L. Rothberg, B. R. Hsieh, R. R. Alfano, *Phys. Rev. B* **49**, 9419 (1994).
19. S. R. Tseng, Y. S. Chen, H. F. Meng, H. C. Lai, C. H. Yeh, S. F. Horng, H. H. Liao, C. S. Hsu, *Synth. Met.* **159**, 137 (2009).
20. N. T. Harrison, D. R. Baigent, I. D. W. Samuel, R. H. Friend, A. C. Grimsdale, S. C. Moratti, A. B. Holmes, *Phys. Rev. B* **53**, 15815 (1996).
21. H. F. Meng, V. C.-H. Chang, *Phys. Rev. B* **60**, 14242 (1999).
22. A. J. Cadby, P. A. Lane, M. Wohlgenannt, C. An, Z. V. Vardeny, D. D. C. Bradley, *Synth. Met.* **111–112**, 515 (2000).
23. U. Scherf, E. J. W. List, *Adv. Mater.* **14**, 477 (2002).
24. D. W. Samuel, R. T. Phillips, *Phys. Rev. B* **52**, R11569 (1995).
25. H. F. Meng, *Phys. Rev. B* **58**, 3888 (1998).
26. E. M. Conwell, J. Perlstein, S. Shaik, *Phys. Rev. B* **54**, R2308 (1996).

27. J. Singh, *Semiconductor Devices: An Introdution*, McGraw-Hill (1994).
28. P. S. Davids, I. H. Campbell, D. L. Smith, *J. Appl. Phys.* **82**, 6319 (1997).
29. M. J. Tsai, H. F. Meng, *J. Appl. Phys.* **97**, 114502 (2005).
30. J. T. Shieh, C. H. Liu, H. F. Meng, S. R. Tseng, Y. C. Chao, S. F. Horng, *J. Appl. Phys.* **107**, 084503 (2010).
31. Y. S. Chen, H. F. Meng, *Phys. Rev. B* **66**, 035202-1 (2002).
32. Z. Shuai, D. Beljonne, R. J. Silbey, J. L. Brédas, *Phys. Rev. Lett.* **84**, 131 (2000).
33. T. M. Hong, H. F. Meng, *Phys. Rev. B* **63**, 075206-1 (2001).
34. M. Wohlgenannt, K. Tandon, S. Mazumdar, S. Ramasesha, Z. V. Vardeny, *Nature* **409**, 494 (2001).
35. M. K. Lee, M. Segal, Z. G. Soos, J. Shinar, M. A. Baldo, *Phys. Rev. Lett.* **94**, 137403-1 (2005).
36. L. C. Lin, H. F. Meng, J. T. Shy, S. F. Horng, L. S. Yu, C. H. Chen, H. H. Liaw, C. C. Huang, K. Y. Peng, S. A. Chen, *Phys. Rev. Lett.* **90**, 036601-1 (2003).
37. H. H. Liao, H. F. Meng, S. F. Horng, J. T. Shy, K. Chen, C. S. Hsu, *Phys. Rev. B* **72**, 113203 (2005).
38. M. A. Baldo, D. F. O'Brien, M. E. Thompson, S. R. Forrest, *Phys. Rev. B* **60**, 14422 (1999).
39. J. Michl, V. Bonacic-Koutecky, *Electronic Aspects of Organic Photochemistry*, Wiley (1990).
40. S. Barth, H. Bässler, *Phys. Rev. Lett.* **79**, 4445 (1997).
41. H. F. Meng, T. M. Hong, *Phys. Rev. B* **61**, 9913 (2000).
42. C. H. Chen, H. F. Meng, *Phys. Rev. B* **64**, 125202-1 (2001).
43. H. Sirringhaus, P. J. Browm, R. H. Friend, M. M. Nielsen, K. Bechgaard, B. M. W. Langeveld-Voss, A. J. H. Spiering, R. A. J. Janssen, E. W. Meijer, P. Herwig, D. M. de Leeuw, *Nature* **401**, 685 (1999).
44. X. M. Jiang, R. Österbacka, C. P. An, Z. V. Vardeny, *Synth. Met.* **137**, 1465 (2003).
45. E. J. Meijer, A. V. G. Mangnus, B.-H. Huisman, G. W. 't Hooft, D. M. de Leeuw, T. M. Klapwijk, *Synth. Met.* **142**, 53 (2004).
46. H. H. Liao, C. M. Yang, C. C. Liu, S. F. Horng, H. F. Meng, J. T. Shy, *J. Appl. Phys.* **103**, 104506 (2008).
47. C. K. Lu, H. F. Meng, *Phys. Rev. B* **75**, 235206 (2007).
48. M. S. A. Abdou, F. P. Orfino, Y. Son, S. Holdcroft, *J. Am. Chem. Soc.* **119**, 4518 (1997).
49. B. Ong, Y. Wu, L. Jiang, P. Liu, K. Murti, *Synth. Met.* **142**, 49 (2004).

50. B. S. Ong, Y. Wu, P. Liu, S. Gardner, *Adv. Mater.* **17**, 1141 (2005).
51. I. McCulloch, M. Heeney, C. Bailey, K. Genevicius, I. Macdonald, M. Shkunov, D. Sparrowe, S. Tierney, R. Wagner, W. Zhang, M. L. Chabinyc, R. J. Kline, M. D. McGehee, M. F. Toney, *Nat. Mater.* **5**, 328 (2006).

Chapter 2

Solution-Processed Organic Light-Emitting Diodes

Hsin-Fei Meng
Institute of Physics, National Chiao Tung University, Hsinchu, Taiwan
meng@mail.nctu.edu.tw

2.1 Introduction

Following the discovery of organic light-emitting diodes (OLEDs) with vacuum-evaporated small molecules in 1987 [1], conjugated polymers were found to exhibit electroluminescence in 1990 [2]. Conjugated polymers are organic semiconductors with bandgap between 1.5 and 3 eV. In earlier works luminescent polymers were shown to be insoluble in common organic solvents and were converted from soluble precursors by thermal curing. Later, orange emitting soluble poly(*p*-phenylene vinylene) (PPV) was developed and shown, in 1993, to have a promising efficiency [3]. Since then, extensive researches have been devoted to polymer light-emitting diodes (PLEDs) by the solution process. The device structure is quite simple. On the transparent conductive ITO glass a conducting polymer, typically poly(3,4-ethylenedioxythiophene) doped with

Polymer Electronics
Edited by Hsin-Fei Meng

ISBN 978-981-4267-84-7 (Hardcover), 978-981-4364-04-1 (eBook)
www.panstanford.com

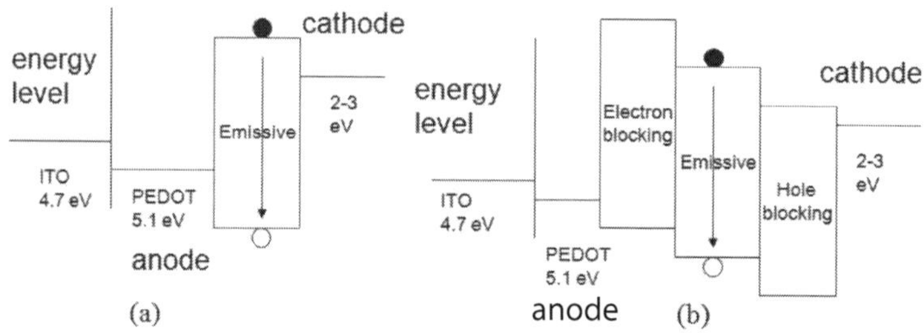

Figure 2.1 (a) In a single-layer PLED, electron and hole current balance is controlled by the transport properties of the emissive material. Work functions of ITO, PEDOT:PSS, and cathode are labeled. Electrons and holes recombine across the bandgap, as indicated by the arrow. (b) In a multi-layer PLED, electron and hole balance is achieved by blocking layers. The electron blocking layer also acts as hole transport layer. The hole blocking layer is also the electron transport layer.

poly(styrenesulfonate) (PEDOT:PSS) from Bayer, is spin-coated in water solution to form a layer of around 40 nm. The purpose is to increase the work function of ITO for easier hole injection and to reduce surface roughness. Electroluminescent conjugated polymers with a thickness of around 80 nm are then deposited on top of the conducting polymer, usually by spin coating. Low work function metals like Ca or Ba are finally evaporated for electron injection. By chemical tuning of the bandgap and carrier mobility of electroluminescent polymers, the emission of PLEDs covers the entire visible spectrum from red to blue. A single-layer PLED device structure is shown in Fig. 2.1a. Electrons are injected from the cathode Fermi level to the conduction band of the polymer and then transported under forward voltage bias to the anode until they meet and recombine with a hole in the valence band. Similarly, holes are injected from the anode Fermi level and then transported toward the cathode. Current will be wasted if an electron goes to the anode without recombining with a hole, or vice versa. There is light emission from the electron–hole recombination in the emissive layer through the exciton state. In conjugated polymers an exciton has an electron–hole binding energy of around 0.3 eV. In an ideal multi-layer device, shown in Fig. 2.1b, carrier blocking layers are added to confine the electrons and holes in the emissive layer and force them to recombine. The device operation principle is discussed in a previous review [4]. Following the initial success of the orange MEH-PPV, PPV co-polymer was shown, in 2000, to have an excellent efficiency for yellow emission [5]. In addition to a high current efficiency close to 10 cd/A, the yellow PPV gives a low turn-on voltage and high brightness of up to 50,000 cd/m^2 below 10 V [6]. The term *super yellow* is used for such PPV derivatives developed by Covion Organic Semiconductor, later acquired by Merck. The chemical structure of super yellow, as an example of PPV family, and the LED performance are shown in Fig. 2.2. In addition to PPV, it has been clear since 2000 that polyfluorene is another class of conjugated polymers with great potential in PLEDs [7]. The chemical structure and optical properties of a typical blue polyfluorene are shown in Fig. 2.3. The polyfluorene derivative covers the blue spectral region, which is impossible for PPV because of an intrinsically low bandgap. Purity and molecular weight distribution are of vital importance for the performance of PLEDs. The metal residues left during synthesis and the chain ends of the polymer may act as carrier traps and exciton-quenching centers. It is relatively easier to synthesize PPV with molecular weight over 500,000, whereas molecular weight over 100,000 is a challenge for polyfluorene. Overcoming these problems, polyfluorene derivatives with metal concentration below 50 ppm and molecular weight approaching 500,000 were synthesized by Dow Chemical. Green PLEDs with a high efficiency of around 10 cd/A were reported in 2004 [8]. Similar results were reported by Cambridge Display Technology in 2007 [9] based on polyfluorene co-polymers. The Ba cathode is used for most of the best results. Ba is, unfortunately, not stable and not very practical for mass production. The exact chemical structures of the polymers in the above two works are not disclosed. Sky blue polyfluorene PLED shows a reasonable efficiency of around 5 cd/A. Red polyfluorene has by far the lowest efficiency: around 2 cd/A. Indeed, photoluminescence quantum efficiency is often reduced as the polymer bandgap decreases. In most polymers with a bandgap below 2 eV, the lowest unoccupied molecular orbital (LUMO) and highest occupied molecular orbital (HOMO) wavefunctions are spatially displaced in two parts in the repeat unit. The wavefunction overlap and quantum optical transition matrix

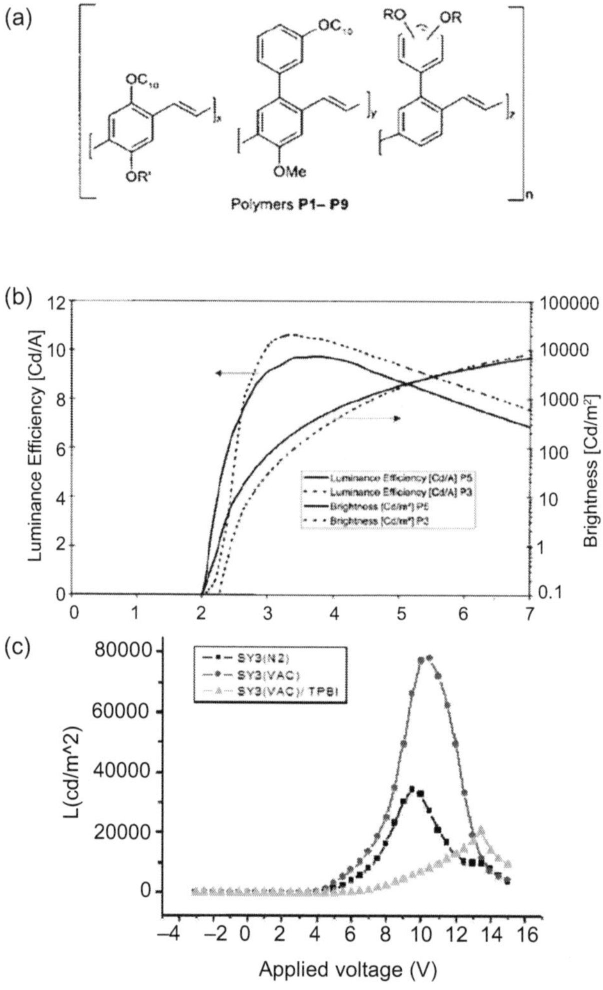

Figure 2.2 (a) Super yellow is a PPV co-polymer. (b) Its efficiency is among the highest for fluorescent PLEDs (data from [5]). (c) Very high luminance is reached ay low voltage (data from [6]).

element are therefore small. For polymers with many repeat units the LUMO is the valence band minimum and HOMO the valence maximum.

In organic electroluminescence, both spin singlet and spin triplet excitons are produced as an electron and a hole capture each other with the ratio 1:3 based on total spin degeneracy multiplicity. In conjugated polymers the triplet excitons do not decay radiatively, i.e., emit photons, to the singlet ground state because of the lack of spin–orbital interaction to break the spin selection rule. The internal quantum efficiency, defined as number of internally emitted photons per electron–hole pair, is therefore limited to below 25%. Total reflection of light at the glass–air interface allows only one-fifth of the light to propagate into open space. The external quantum efficiency for those polymers thus has an ultimate upper bound of 5%. Electroluminescence with singlet excitons is classified as fluorescence OLED and electroluminescence with triplet excitons is classified as phosphorescence OLED. For yellow, green, and sky blue fluorescence PLEDs discussed above, the external quantum efficiency is already close to the theoretical limit of 5%. There is therefore not much room for further improvement. In vacuum-deposited OLEDs, the iridium metal–organic complex molecule is shown to be an excellent phosphorescent emitter as it is doped into an organic host [10].

The radiative quantum transition from triplet excitons to the ground state is now slightly allowed due to the spin–orbital interaction of the heavy metal of iridium. Inspired by the success of the vacuum-deposited OLED, the iridium complex was introduced, around 2004, in a polymer host for the PLED [11]. An efficiency over 40 cd/A was achieved for green; this is far higher than the 10 cd/A ever reached for conjugated polymers alone. Because of the rapid development of phosphorescent green and red PLEDs, they appear to be more promising for future applications than PPV and polyfluorene. For blue and white emission, phosphorescent PLEDs suffer from limited lifetime and efficiency. Therefore both fluorescence and phosphorescence need to be taken into account for applications. In this chapter we will review the best results so far for blue and white PLEDs, including both fluorescence and phosphorescence. For green and red we will cover only phosphorescence because of the higher efficiency. In many device designs both polymers and small molecules are used. We therefore sometimes prefer the term solution OLED to polymer LED. Vacuum deposition and solution deposition are mixed and combine some OLED fabrication. Because the unique advantage of large-area low-cost coating of solution OLEDs will be greatly compromised by the introduction of any layer by vacuum deposition, in this chapter we will discuss mostly devices fabricated entirely by solution except for the metal cathode.

2.2 Blue Fluorescence

2.2.1 *Polymers*

Sky blue polyfluorene co-polymers with emission peak around 470 nm are developed by Dow Chemical and other companies. Unfortunately the chemical structure is not open to the scientific community. Because of the demanding requirement on molecular weight and impurity level, the polyfourene co-polymers synthesized in academic institutions with open chemical structures in general have a lower performance than the one achieved in industry. The blue polyfluorene has ionization potential in the range of 5.5–5.8 eV with a bandgap of around 2.8 eV [12]. It has a high photoluminescence quantum efficiency in the solid state, so the PLED efficiency is mostly determined by carrier injection and transport. Hole and electron mobilities can be measured by time-of-flight [13], space-charge-limited current fitting [13], and transient electroluminescence [14]. Most of the polyfluorene film has hole mobility in the range of 10^{-5} cm^2/Vs. High electron mobility is required for a balance electron–hole current and high efficiency. The sky blue polymer BP105 developed by Dow Chemical has a relatively high electron mobility of over 10^{-5} cm^2/Vs [14] and shows an efficiency of 2.7 cd/A for single-layer PLEDs, where the emissive layer is sandwiched between the conductive polymer PEDOT:PSS anode and the LiF/Ca/Al cathode. In such a single-layer structure there is a large hole injection energy barrier in the anode. The PEDOT:PSS has a work function of around 5.1 eV, while the ionization potential of blue polyfluorene is at least 5.5 eV. The energy barrier of over 0.4 eV makes it difficult to supply enough holes in combination with the electrons. In order to improve the PLED efficiency a hole transport layer with ionization potential between the anode work function and the emissive polymer ionization

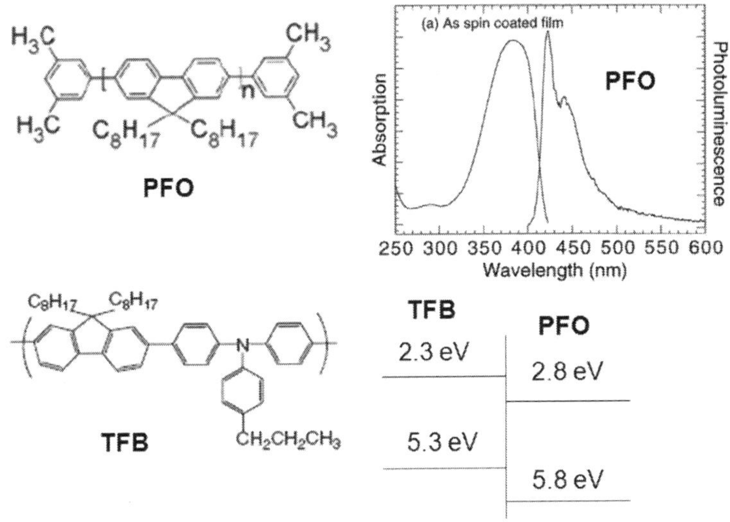

Figure 2.3 PFO is the archetypical polymer with fluorene repeat unit and blue emission. The polyfluorene derivatives cover the entire visible spectrum. The hole transport polymer TFB is often used to improve the hole injection from the PEDOT:PSS anode. The LUMO and HOMO levels of TFB and PFO are shown. Data from M. Ariu *et al.*, *Synthetic Metals*, 111–112, **607** (2000).

potential is inserted. The holes can then be injected into the hole transport material first with a lower barrier, and then injected into the emissive layer. Co-polymers with tri-phenyl amine and fluorine units are the most common hole transport material because of the appropriate ionization potential of around 5.3 eV and good hole mobility of around 10^{-4} cm^2/Vs. One important example is TFB, shown in Fig. 2.3. Besides hole transport, the interlayer also serves as electron blocking layer which prevents electrons from entering the anode and become wasted. Small electron affinity is necessary for electron blocking. Indeed the electron affinity of tri-phenyl amine is around 2.4 eV, far smaller than the typical 3.0 to 2.7 eV for the sky blue polyfluorene. In practice, the hole transport material and the emissive polymer usually share the same solvents. The latter therefore cannot be simply spin-coated on the former without causing dissolution. Various strategies are used for the fabrication of the hole transport layer. After thermal annealing up to 180°C it is found that a thin layer of TFB with a thickness of less than 5 nm becomes insoluble, perhaps due to the entanglement with the underlying conducting polymer for the anode. The soluble part of the TFB is then removed by rinsing with pure solvent in a spinner.

The method is called spin-rinsing and has proved to be widely successful despite the very small thickness after rinsing. Bi-layer device with the structure TFB/BP105 shows an efficiency of 3.5 cd/A by spin-rinse [15]. Thermal annealing up to 180°C for a hole transport polymer PFB is shown to have some resistance to the solvent xylene [16]. A greenish blue PLED with an efficiency of up to 10 cd/A is achieved in the bi-layer device with annealed PFB, which is only partially dissolved by the weak solvent of xylene. Only a low luminance of 200 cd/m^2 is reached. Both spin-rinse and the weak solvent method cannot produce a well-defined bi-layer structure with arbitrary thickness. In 2006 a buffer liquid method was proposed to make general multi-layer polymer structures [17]. In this method the first polymer layer is covered by a buffer liquid of high viscosity. The second layer is then spin-coated on top of the buffer liquid. Most of the buffer liquid is spun away by the centrifugal force, and subsequently the small amount of residue is removed by baking. The buffer liquid therefore needs to satisfy three conditions: it must be non-solvent for the first layer, have a high viscosity to withstand the solvent of the second layer, and evaporate at a relatively low temperature. Because of the hydrogen bond the small molecule 1,2-propylene glycol meets all the three conditions. The buffer liquid method and the resulting bi-layer device structure are shown in Fig. 2.4. The hole–transport polymer TFB with a well-defined thickness of 30 nm is introduced to the sky blue PLED with BP105 as the emissive polymer. The high efficiency of 9.1 cd/A at 6 V and high luminance of 26,000 cd/m^2 at 8 V are realized. Apparently the hole transport layer with a thickness of around 20–40 nm is more effective in hole injection and electron blocking than a mere 5 nm electron-blocking layer made by spin-rinsing. In fact, there is probably no electron blocking effect at all for spin-rinsing as the thin layer may not form a uniform film, so the electrons can easily leak into the anode through the openings of the hole transport layer. The 9.1 cd/A efficiency above

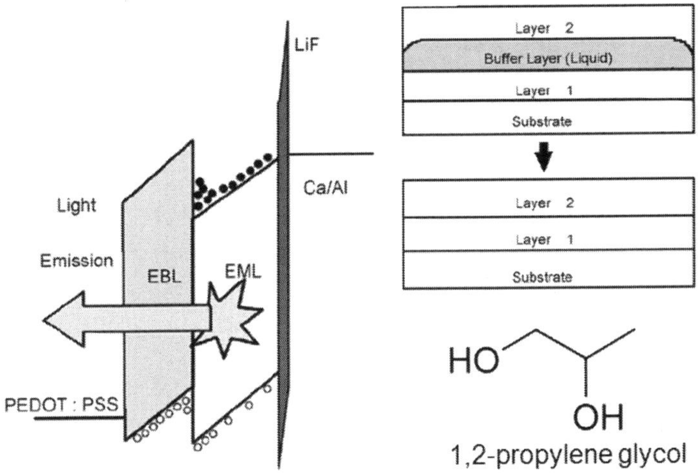

Figure 2.4 Bi-layer blue PLED is made by the liquid buffer of 1,2-propylene glycol. The high viscosity of the liquid buffer prevents the dissolution of layer 1.

corresponds to an external quantum efficiency of 5.5%, which is about the theoretical limit for fluorescence PLED. Ohmic contact for electron injection is crucial for the cathode. For pure metals, barium with work function as low as 2.5 eV is the only element which forms the desired Ohmic contact. Because of the high bandgap even calcium with work function of 2.9 eV is not low enough. Barium is, however, highly reactive to air. A composite cathode involving alkaline metal salt is commonly used, including LiF/Ca/Al and Cs_2CO_3/Al [18]. Photo-emission experiments demonstrated that some of the alkaline metal atoms dissociate with the halogen partner and donate their valence electrons to the emissive polymer. The intrinsic conjugated polymer therefore becomes heavily n-doped near the cathode interface where an Ohmic contact forms. Unfortunately highly reactive elements like calcium and cesium are still involved in these composite cathodes. The search for a stable cathode remains a challenge for PLEDs.

While the sky blue PLED has reached the theoretical limit, the deep blue PLED with emission peak around 430 nm proves to be more difficult. Few polyfluorene derivatives have a high enough bandgap for deep blue emission. In fact the archetypical PFO shown in Fig. 2.3 with ionization potential 5.8 eV and electron affinity 2.8 eV appears to be the only main-chain structure for deep blue polymers. Despite the high solid-state photoluminescence quantum efficiency of around 50%, PLEDs made of PFOs usually show poor efficiency mostly due to the difficulty in hole injection. Furthermore, the PFO emission spectrum is extremely sensitive to the ketone defect with the carbon–oxygen double bond [12]. Such defect rises often at the bridge non-conjugated carbon atom of the fluorene unit with only one alkyl chain during device operation. Indeed the originally deep blue emission turns into a green emission caused by the ketone defect some time after the voltage bias. Such color instability can be improved by the spiral chemical structure in the fluorene unit. An efficiency of 3 cd/A is realized for the spiro-polyfluorene by Covion, but the detailed chemical structure is not disclosed [19]. Similar to the sky blue polymer, a hole transport layer of TFB is useful to improve the hole injection and electron blocking. The buffer liquid method is employed to fabricate a bi-layer device with the structure TFB/PFO and the efficiency is raised from 1.2 cd/A for the single-layer to 2.7 cd/A for the bi-layer device [20]. As discussed above, a high molecular weight is crucial for high electroluminescence efficiency because chain ends form defect levels. It turns out to be difficult to synthesize PFO derivatives with molecular weight more than 100,000 as the molecular weight distribution depends on the details of the synthesis conditions and varies from batch to batch. PFO with molecular weight over 360,000 shows an enhanced efficiency of 3.8 cd/A, corresponding to 3.7% of external quantum efficiency [21]. So far, no deep blue polymers with high efficiency and good material reproducibility are available.

2.2.2 *Small Molecules*

Conventionally organic light-emitting diodes are divided into two categories: polymer LEDs processed by solution and small-molecule LEDs processed by vacuum deposition. The line between them is in fact not absolute. Indeed polymers can never evaporate, because they decompose as the temperature rises. On the other hand, there exist small molecules that can be deposited by spin coating or other ways in solution. The problems for most of the small molecules are that first they do not have enough solubility and second they do not have enough interaction among one another to form a uniform film like polymers. There are entanglements among the long polymer chains. Such interaction is crucial during spin coating to prevent breaking of the film into pieces. Because of the absence of entanglement small molecules in solution are like sands in water and tend to scatter into pieces during spin coating. In order to make it possible for small molecules to be processed in solution, chemical modifications are employed. One way is to add some alkyl side chains to the small molecule without changing the main conjugated part. The side chain improves both the solubility and the interaction among the molecules. It also makes the films more amorphous by reducing the tendency to crystallize. Unlike polymers, most conjugated small molecules have a strong tendency to crystallize, especially during the drying process of the solvent. Crystallization destroys the uniformity and causes breakdown of the LED. Even though the introduction of the alkyl chain usually does not alter the electronic levels, it does affect the mobility as carrier transport is hindered by the insulating side chains. The solubility and mobility are therefore two contradictory factors to be taken into account in molecular design.

In addition to side chain, another way to process small molecules in solution is oligomer, which is basically a very short polymer with only a few repeat units. Like polymers, oligomers contains the side chains to ensure solubility. Their relatively large size and the side chains give enough molecular interaction to form a uniform film. The electronic levels are similar to the parent polymer with infinite repeat units, but the oligomers allow more room for electronic tuning by the end groups. Another great advantage of oligomers over polymers is that the former are much better to purify and have no complications from molecular weight distribution. Two works on high-efficiency blue OLEDs processed in solution using alkyl chain and oligomer approaches respectively are discussed below.

A tri-layer small-molecule blue OLED is made by blade coating to prevent dissolution. Through rapid moving of the blade followed immediately by heating to expel the solvent, it was demonstrated that blade coating can be used to deposit multiple layers of organic

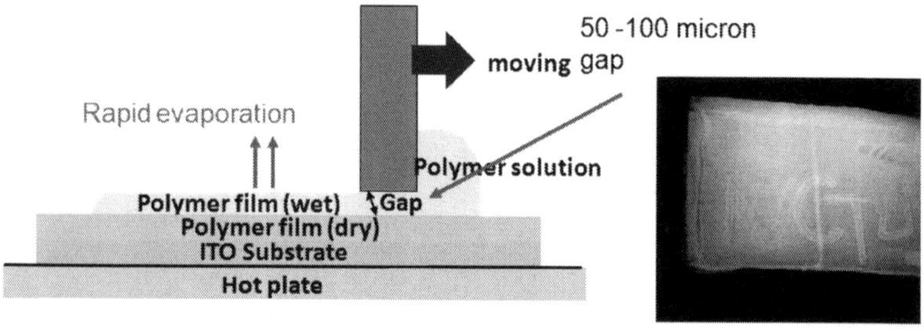

Figure 2.5 Multiple layers of organic semiconductors can be made by blade coating. Rapid solvent evaporation by substrate heating avoids dissolution. In the picture a pattern of NCTU is made on the first yellow polymer. The second blue polymer deposited by blade coating does not destroy the pattern. Data from [22]. See also Color Insert.

semiconductors sharing a common solvent [22]. The idea of blade coating is shown in Fig. 2.5. In addition to multi-layer deposition, blade coating has almost no material waste in sharp contrast to spin coating, where more than 90% of the materials are spun away. The device includes a hole transport layer, emissive layer, and electron transport layer (ETL). The ETL also plays the role of hole blocking. Such a structure is typical for vacuum-deposited small-molecule OLEDs but was difficult to make before the invention of the blade-coating technique. For the hole transport layer the small-molecule NPB is modified to be DMFL-NPB, as shown in Fig. 2.6. For the emissive layer a host–guest system is used. Both the host and guest are blue-emitting small molecules containing alky side chains. The host has a slightly larger bandgap than the guest. Such strategy is often used for vacuum-deposited blue OLEDs for improved efficiency. The possible reason is that the wide bandgap of the blue emitter makes its excitons susceptible to defect levels in the energy gap which act as non-radiative recombination centers. In the homogeneous case the excitons diffuse freely and are likely to encounter a defect level. In the host–guest system the excitons are all transferred to the guest and become localized because they form a potential well in the exciton energy profile. Such localization prevents the non-radiative recombination due to the defect. Small-molecule TPBI is used for the electron transport layer. Because of the large size, TBPI, even without modification, has enough solubility

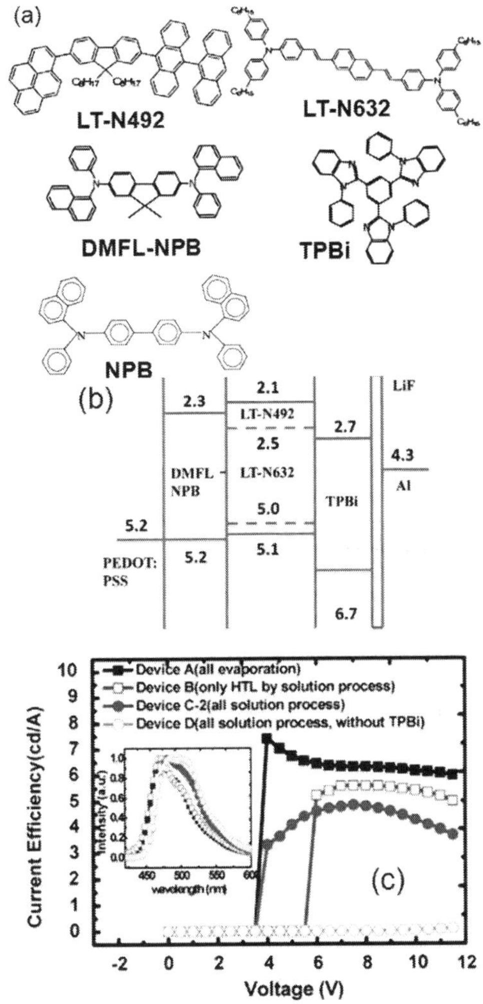

Figure 2.6 (a) The blue emitters LT-N492 and 632 have side chains to improve the solubility. DMFL-NPB is modified version of hole transport material NBP for better solubiity. TPBI is the typical electron transport material. (b) The energy levels of the blue OLED is shown. (c) The blue line is for tri-layer device made by blade coating. Black line is for same device made by vacuum depositon. Reprinted from *Org. Electron*, 10, J. D. You, S. R. Tseng, H. F. Meng, F. W. Yen, I. F. Lin, S. F. Horng, All-solution-processed blue small molecular organic light-emitting diodes with multi-layer device structure, 1610. Copyright 2011, with permission from Elsevier. See also Color Insert.

and molecular interaction to form uniform films. The solution-processed tri-layer OLED has blue emission peaked at 470 nm and efficiency of 5 cd/A [23]. The same structure can be made entirely by vacuum deposition as only small molecules are involved. Such a vacuum-deposited device shows an efficiency of 7 cd/A. The chemical structures of the materials and the device performance are shown in Fig. 2.6. The slightly lower efficiency by blade coating may come from some level of dissolution and the morphology of TBPI. Indeed the annealing temperature of TPBI must be kept lower than 50 degrees. Even at such a low temperature some aggregates of the TBPI molecules are still clear in the atomic force microscope images.

A high-efficiency blue OLED is demonstrated using oligomers as the emitter [24] with up to three fluorene repeat units. The electronic levels are tuned by the CN end groups. Unlike most blue emitters the fluorene oligomers have an ionization potential of only 5.2 eV, enabling efficient hole injection. No green emission results as in the case of PFO polymer due to high purity which prevents the formation of ketone defects. When the oligomer is added to the CBP host, deep blue emission peaked at 439 nm with an efficiency of 6 cd/A is realized, corresponding to 8% external quantum efficiency. The efficiency drops to 4% at high luminance. The host–guest system shows better performance than the homogeneous emissive layer probably due to exciton localization. Even though the emissive layer is solution processed, TBPI is still evaporated as the electron transport layer. The great potential of oligomer over polymer for a solution OLED is, however, clearly demonstrated in this example.

2.3 Blue and White Phosphorescence

A stable deep blue phosphorescent metal complex is difficult to develop. For sky blue the iridium complex FIrpic shown in Fig. 2.7 is by far the most popular. FIrpic was originally developed for vacuum deposition but it does have some solubility. In vacuum-deposited OLEDs the electron–hole current balance is achieved by electron and hole blocking layers which confine the carriers in the emissive layer and force them to recombine. Because of the dissolution by spin coating, solution-processed phosphorescent

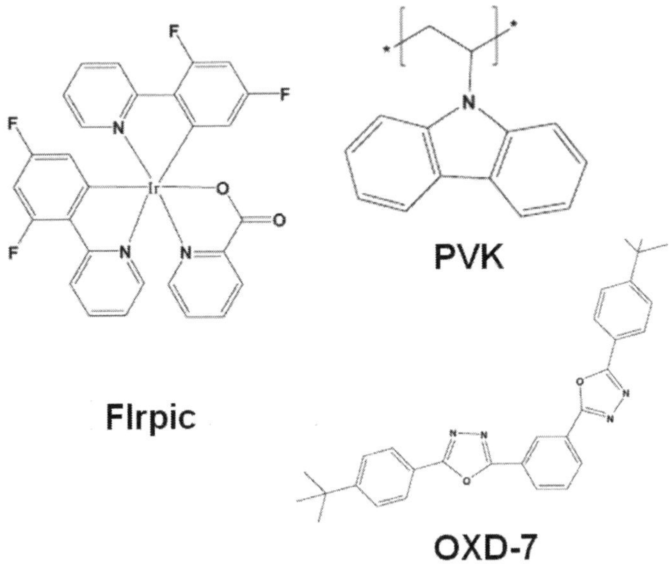

Figure 2.7 The blend of PVK and OXD-7 is used as the host for a sky blue phosphorescent PLED. FIrpic is the sky blue emitter.

OLEDs achieve electron–hole balance by mixing electron and hole transport molecules in the same layer. The electron and hole current can be fine-tuned by the relative concentration of the electron transport molecules in the blend. FIrpic is used as the triplet sky blue emitter in a host composed of hole-transporting PVK and electron-transporting OXD-7. The chemical structures are shown in Fig. 2.7. Both the current density and device efficiency are very sensitive to OXD-7 concentration, indicating that the electron–hole balance occurs only at optimal concentration. In phosphorescent OLEDs the triplet exciton energy of the host must be higher than that of the emitter, otherwise the triplet exciton formed in the emitter will be transferred to the host and decay nonradiatively. Both PVK and OXD-7 have a higher triplet exciton energy than FIrpic, but it may not be high enough for deep blue emitters. In these devices CsF/Al is used for effective electron injection. Such devices were reported to have a high efficiency of 22 cd/A by Osram [25]. Such an excellent result is yet to be reproduced by other groups. FIrpic is, however, known

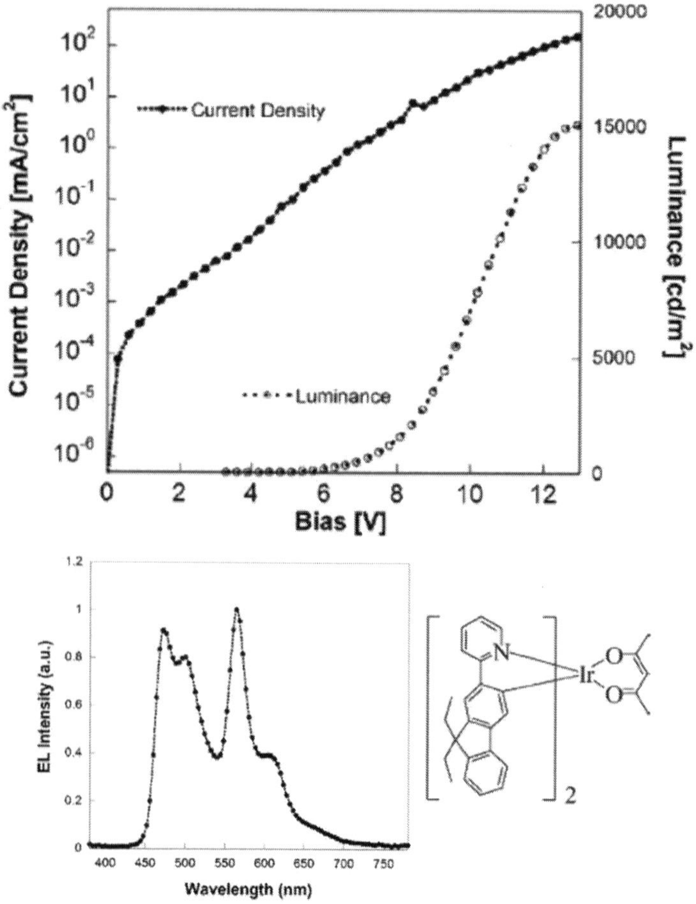

Figure 2.8 A white phosphorescent OLED is made with the yellow emitter shown and sky blue FIrpic. An efficiency of over 40 cd/A is achieved. Data from [27].

to be photochemically unstable because the fluorine atom acts as an electron acceptor under excitons and causes further chemical reaction. There is some uncertainty on lifetime.

On the basis of the sky blue device concept, white phosphorescent PLEDs can be made by adding a yellow triplet emitter to the polymer host. The osmium complex is added to a blend containing PVK, OXD-7, and FIrpic to produce white emission [26].

Electron injection improves by adding a water-soluble polyfluorene derivative as the electron transport layer between the emissive layer and the Ba cathode. The dipole in the polar side chain of the polyfluorene reduces the injection barrier for the electrons. Li_2CO_3 is used to *n*-dope the electron transport layer such that the conductivity is enhanced and the driving voltage of OLEDs is reduced. An efficiency of 36 cd/A is achieved for this solution-processed white PLED with an *n*-doped electron transport layer [26]. One concern for this device is that water and alcohol have poor wetting property on top of the emissive layer. Reproducibility and uniformity for large-area spin coating may be a problem. High-efficiency white PLEDs can be realized even without the electron transport layer. Yellow iridium complexes containing a carbazole unit are added to the host blend of PVK, OXD-7, and FIrpic. An efficiency of 43 cd/A is demonstrated [27]. The chemical structures and device performance are shown in Fig. 2.8. Once again Ba is used as the cathode. Despite the high efficiency of the white devices discussed above, the unstable Ba cathode causes uncertainty in mass production.

2.4 Green Phosphorescence

Green iridium complexes are successfully introduced to a polymer matrix as the emissive layer as in the case of the sky blue iridium complex FIrpic. $Ir(ppy)_3$ is a green triplet emitter with a rapid radiative transition from its triplet exciton state to the ground state. For small-molecule OLEDs by vacuum evaporation, green emission with a nearly ideal 20% external quantum efficiency is achieved using $Ir(ppy)_3$ as the emitter doped in the CBP host [10]. $Ir(ppy)_3$ itself has poor solubility; the methyl group is therefore added to the ligands to improve the solubility, resulting in $Ir(mppy)_3$ shown in Fig. 2.9. Hole transport, electron transport, and the soluble green emitter form a blend as the emissive layer. The electron and hole currents are fine-tuned by their ratio as in the case of sky blue. PVK is commonly used as the hole transport host, PBD as the electron transport material, and $Ir(mppy)_3$ as the emitter. The hole injection and transport is further enhanced by adding the small-molecule TPD shown in Fig. 2.9. The 5.8 eV ionization potential of PVK makes it difficult for holes to be injected from the conducting polymer anode PEDOT:PSS with a work function of 5.1 eV. TPD is a common material for vacuum-deposited OLEDs and has an ionization potential of 5.5 eV and good hole transport mobility. Note that TPD and PBD have triplet exciton levels high enough to confine the triplet excitation in the green iridium complex but not so for the sky blue iridium complex FIrpic discussed above. That is why PBD is replaced by OXD-7, while TPD cannot be used for FIrpic. For optimal electron–hole balance the ratio of the materials is 61% of PVK, 24% of PBD, 9% of TPD, and 6% of $Ir(mppy)_3$. The sensitivity to the ratio makes the process of the devices not very reproducible. There is no such problem for vacuum-deposited OLEDs as the electron–hole balance is forced in the multi-layer structure and the emitter is in a pure CBP host. As in the case of fluorescent PLEDs, hole injection can be improved by adding a hole transport layer between the anode and the emissive layer. Spin-rinse is used to add a layer

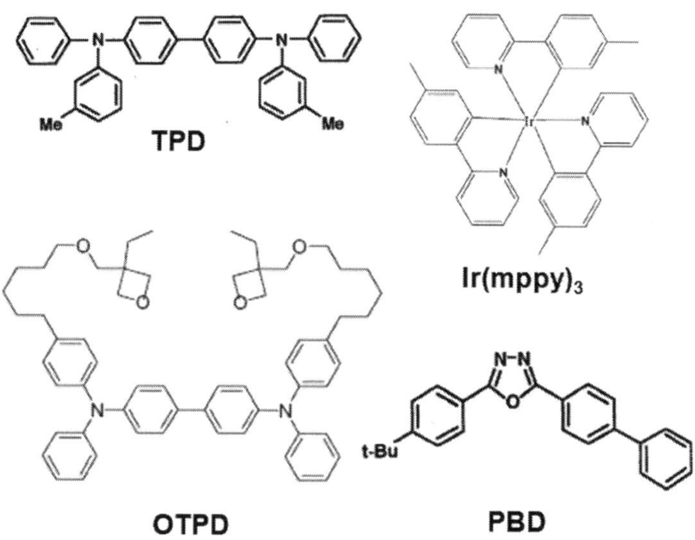

Figure 2.9 In green phosphorescent PLEDs the cross-linking hole transport material OTPD is modified from TPD. PBD and TPD are added into PVK to form the host. $Ir(mppy)_3$ is the green triplet emitter. It has a methyl side chain to improve solubility.

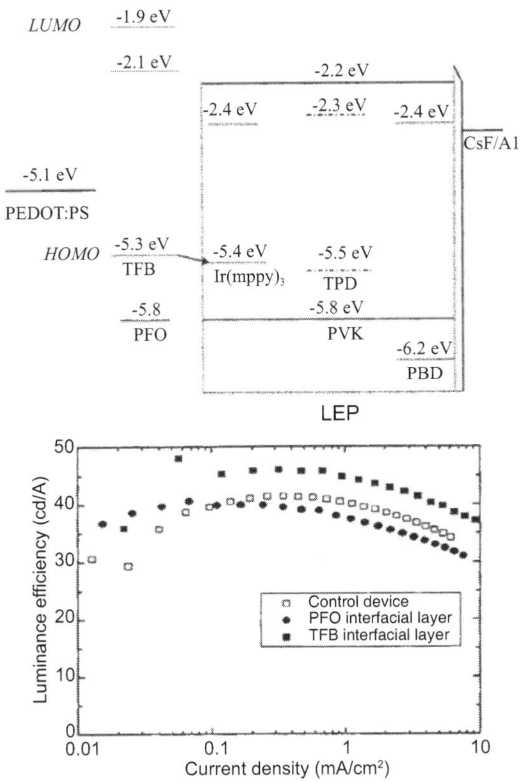

Figure 2.10 The energy levels and the device performance of green phosphorescent PLEDs is shown. The thin TFB layer improves the hole injection. There is some controversy on the electron affinity of polyfluorene because electrochemical measurement and optical gap give different values. Data from [28].

of TFB polymer a few nanometers thick. An efficiency of 40 cd/A is achieved with this thin TFB layer [28]. The chemical structures are shown in Fig. 2.9. The device structure and performance are shown in Fig. 2.10. The performance can be further improved by using a 20 nm thick cross-linked hole transport layer [29]. Cross-linking function groups are added to the hole transport molecule TPD. Under ultraviolet curing the function groups react with each other to form a three-dimensional network. The cured film is no longer composed of isolated small molecules and becomes insoluble for organic solvents. An excellent efficiency of 67 cd/A is achieved using the cross-linked hole transport layer in combination with the blend of PVK, OXD-7, PBD, and Ir(mppy)$_3$ as the emissive layer [29]. The corresponding external quantum efficiency is 18%, approaching the theoretical limit of 20% as in the case of evaporated OLEDs. Indeed the green phosphorescent device based on Ir(mppy)$_3$ is the most efficient among all solution-processed OLEDs. The reason that a relatively thick layer of cross-linked hole transport layer gives a higher efficiency than the ultrathin layer made by spin-rinsing is that in addition to helping hole injection the former has the effect of electron blocking while the latter does not. As discussed above, the ultrathin TFB layer by spin-rinsing is not expected to be uniform and electrons in the emissive layer can leak through into the anode.

In all the green phosphorescent devices discussed above, the CsF/Al cathode is used for easy electron injection. As discussed above, through the reaction with aluminum some cesium atom is dissociated from the fluorine atom and donates its valence electron to the organic materials. Cesium has a low work function of 2.1 eV and forms a conductive doping of the emissive layer. CsF is, however, difficult to handle in the device fabrication because it has a strong tendency to absorb water from the surroundings. It becomes a soft substance once exposed to moisture. CsF therefore causes some instability of the vacuum chamber if the moisture level is not under stringent control. The presence of CsF in the device also puts a strict requirement on the encapsulation and reduces the lifetime. In contrast to CsF, LiF/Al has proved to be a stable cathode in vacuum-deposited OLEDs, and a lifetime of tens of thousands of hours has been widely demonstrated. In spite of the high efficiency, the mass production of phosphorescent PLEDs will meet great challenges if the CsF cathode has to be used. In general the LiF/Al cathode does not have a low enough work function to effectively inject electrons in the emissive layer. An electron transport layer with an electron affinity higher than that of the emissive layer is required in between as in the case of vacuum-deposited.

OLEDs. The common electron transport materials are, however, small molecules developed for vacuum deposition. Nevertheless the electron transport molecules TBPI and TAZ shown in Fig. 2.11 do have some level of solubility and a thin film can be formed by blade

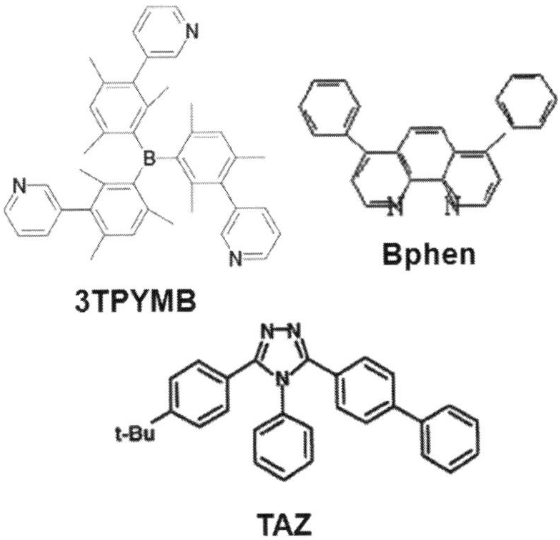

Figure 2.11 Electron transport small molecules that are solution processed on the emissive layer for green phosphorescence. An efficiency ranging from 30 cd/A to 40 cd/A can be achieved. LiF/Al is used as the cathode.

coating. Multi-layer green phosphorescent devices can be made by blade coating with TFB as the electron transport layer, the blend including $Ir(mppy)_3$ as the emissive layer and TBPI or TAZ as the electron transport layer. Finally LiF/Al is evaporated as the cathode. An efficiency of 40 cd/A can be achieved [30]. The possibility to replace the unstable CsF with the relatively stable LiF removes one major obstacle for the mass production of PLEDs.

In order to be used as emitters in solution-processed OLEDs, self-quenching of the iridium complex molecules must be avoided. Many luminescent molecules have a high photo-luminescence efficiency in solution but low efficiency in solid state. This is because the wavefunctions are altered as two or more molecules are in close contact. The symmetry of the wavefunctions may change and an allowed quantum transition becomes forbidden. The LED performance will be poor if self-quenching, or concentration quenching, occurs in the emissive layer. The iridium complexes therefore need to be uniformly dispersed in the polymer host in order to avoid aggregates. The concept of dendrimer is introduced to reduce the self-quenching of the emitter. The radiative transition comes from

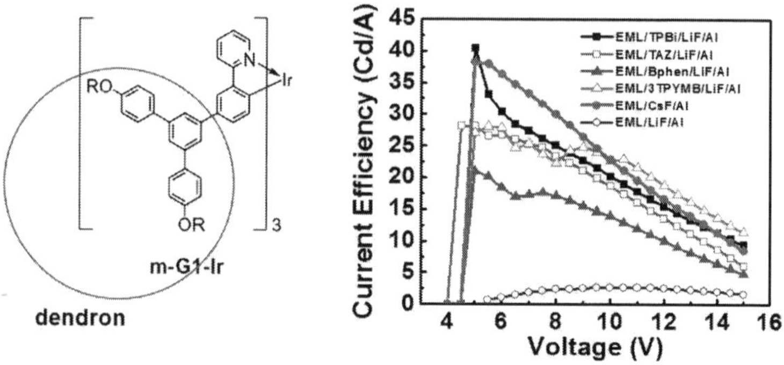

Figure 2.12 The green dendrimer has an iridium complex core and surrounding bulky dendrons to protect the core from self-quenching. The ETL is deposited in solution by blade coating. A luminance of around 10,000 cd/m^2 is reached at 10 V. Reprinted from *Org. Electron.*, **11**, L. C. Ko, T. Y. Liu, C. Y. Chen, C. L. Yeh, S. R. Tseng, Y. C. Chao, H. F. Meng, S. C. Lo, P. L. Burn, S. F. Horng, Multi-layer organic light-emitting diodes processed from solution using phosphorescent dendrimers in a polymer host, 1005. Copyright 2011, with permission from Elsevier.

the transition of an electron from the LUMO to the HOMO level. For the common phosphorescent emitter the HOMO and LUMO wavefunctions lie at the transitional metal ion core and the ligand. Bulky groups called dendrons are added to the ligand to protect the core including the metal and the ligand as shown in Fig. 2.12. The solid-state photoluminescence quantum efficiency is improved in dendrimers as the dendrons make the HOMO and LUMO levels intact even when the molecules are in close contact. The dendrons also further increase the solubility. Consider the example of the $Ir(ppy)_3$ core; the modification $Ir(mppy)_3$ with the methyl group still has a relatively low solubility compared with common polymers. On the contrary, the modification of dendrimer m-G1-Ir, indicating generation one in meta position, shown in Fig. 2.12, makes the molecule as soluble as most polymers. The solubility is important on several aspects. First it gives uniform distribution of the emitters in the polymer host to avoid host emission. Second, it makes the accurate concentration control of the blend easier. Finally, it allows the use of weak solvents like toluene rather than strong solvents like chlorobenzene. The weak solvent makes multi-layer blade coating easier and is far less toxic than chlorobenzene. High-efficiency OLEDs are achieved by mixing green dendrimers in the CBP or TCTA host [31]. An efficiency of 40 cd/A is reached using a vacuum-evaporated TPBI electron transport layer. All solution-processed dendrimer OLEDs are made with a blend of PVK, TPD, and PBD as the emissive layer. The electron transport layer of TBPI, TAZ, and 3TBYMB, shown in Fig. 2.11, is deposited by blade coating. LiF/Al is used as the cathode. The efficiency is between 30 and 40 cd/A as shown in Fig. 2.12 [32]. Unlike $Ir(mppy)_3$, whose concentration in the blend cannot be raised over 6%, dendrimer concentration can be raised to be over 50%. High emitter concentration has the potential to reduce the driving voltage. As the emitter concentration is lower than the percolation threshold the carrier cannot be transported through the emitter. For high concentration the carrier can move directly through the emitter pathway and a lower voltage is required for a given operation current density. Despite the advantages, the synthesis of the dendrimer is relatively complicated.

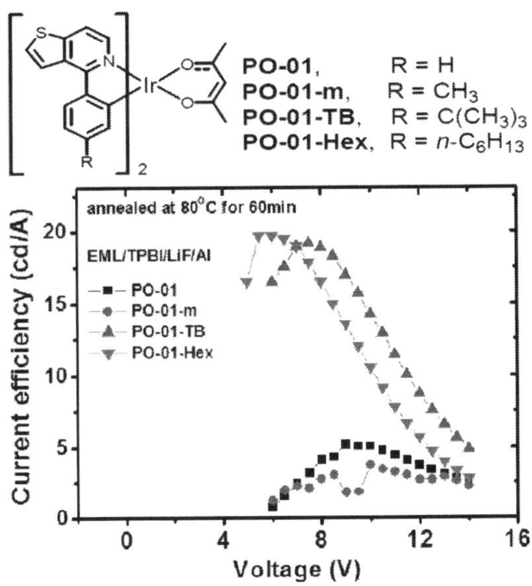

Figure 2.13 Four orange iridium complexes with emission peak at 560 nm are compared. They share the same core but differ in side groups. TB and Hex side groups give good solubility and high device efficiency.

2.5 Red Phosphorescence

Similar to the case of sky blue and green phosphorescent PLEDs, the red emitting iridium complex can be added to a polymer host as the emissive layer. Uniform dispersion of the red emitter appears to be more critical than the sky blue and green ones because the small bandgap of the red emitters makes them deep traps for carrier transport. So the carriers have to hop through the emitter molecules. If they are separated too much, hopping transport will be difficult. In order to demonstrate the effect of dispersion, a series of orange emitters with an identical core but different side groups shown in Fig. 2.13 are compared for device performance. PO-01 has no side chain and PO-01-m has only a short methyl chain, so they have poor solubility and a poor efficiency of around 5 cd/A in the blend host of PVK, PBD, and TPD. On the contrary, PO-01-TB and PO-01-Hex have more carbon atoms in the side chain to enhance the solubility, so they have a much higher efficiency of around 20 cd/A in the same host [33]. The current density of the soluble emitters PO-01-TB and PO-01-Hex is also much higher, indicating the ease of hopping transport due to the uniform distribution of the emitters in the host. The electronic levels of all the four molecules are the same, so the difference comes entirely from the dispersion in the polymer host.

Many red iridium complexes with functional groups to enhance solubility have been synthesized and shown to have reasonable LED efficiency in polymer host with a vacuum-deposited electron transport layer [34–36]. In particular, 17.5 cd/A corresponding to 10% external quantum efficiency is demonstrated using the iridium complex (Et-Cvz-PhQ)2Ir(pic) in a host containing PVK, TPD, and OXD-7. The electron transport layer of OXD-7 is vacuum deposited and Ba is used as the cathode [38]. All solution-processed red phosphorescent PLEDs are made using Ir(piq)2acac as the emitter in a polymer host. Cross-linked TPD is used as the hole transport layer and CsF as the cathode. An efficiency of 9.7 cd/A, corresponding to 10% external quantum efficiency, is achieved [29]. Ir(piq)2acac, however, has only moderate solubility, so process reproducibility is a concern. The iridium complex Ir(DMFPQ)2acac is used as

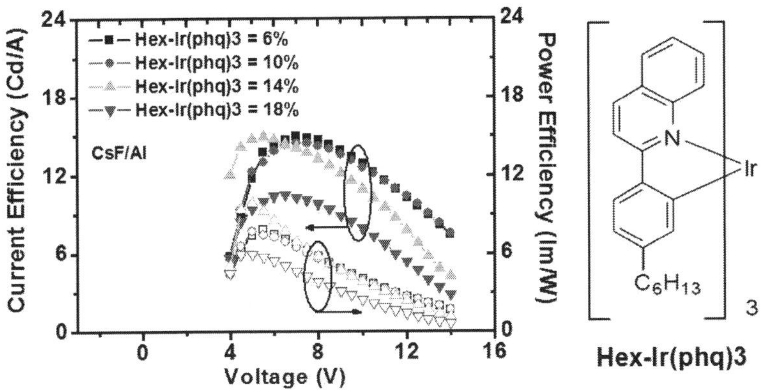

Figure 2.14 Six carbon side chains are added to the orange red iridium complex with emission peak at 600 nm. High efficiency is achieved due to the uniform dispersion of the emitter in polymer host.

the emitter in a host blend of PFO and PBD. Electron injection is enhanced by adding the water-soluble polymer PFN between the emissive layer and the Ba cathode. An efficiency of 9.8 cd/A, corresponding to 18% external quantum efficiency, is reported [37]. An efficiency of 17 cd/A, corresponding to external quantum efficiency of 10%, can be achieved by adding a linear six-carbon side chain to the orange red emitter $Ir(phq)_3$ used for vacuum deposition as shown in Fig. 2.14 [39]. As discussed above, both CsF and Ba are unstable cathodes and LiF is preferred for practical production. TPBI deposited by solution is added as the electron transport layer in combination with the LiF/Al cathode for devices with Hex-$Ir(phq)_3$ emitters. Even though the efficiency is about the same as that of the CsF/Al cathode, the luminance is much lower for a given voltage. There is no such major difference for green phosphorescent devices, suggesting that the deep trap nature of the red emitters in the host puts a high demand on the electron injection contact in order to supply enough electron current for high recombination and luminance.

References

1. C. W. Tang, S. A. VanSlyke, *Appl. Phys. Lett.* **51**, 913 (1987).
2. J. H. Burroughes, D. D. C. Bradley, A. R. Brown, R. N. Marks, K. MacKay, R. H. Friend, P. L. Burn, A. B. Holmes, *Nature*, **347**, 539 (1990).
3. G. Yu, C. Zhang, A. J. Heeger, *Appl. Phys. Lett.* **64**, 1540 (1993).
4. R. H. Friend, R. W. Gymer, A. B. Holmes, J. H. Burroughes, R. N. Marks, C. Taliani, D. D. C. Bradley, D. A. Dos Santos, J. L. Bre'das, M. Logdlund, W. R. Salaneck, *Nature*, **397**, 121 (1999).
5. H. Becker, H. Spreitzer, W. Kreuder, E. Kluge, H. Schenk, I. Parker, Y. Cao, *Adv. Mater*, **12**, 42 (2000).
6. C. S. Hsu, H. F. Meng (unpublished results).
7. M. T. Bernius, M. Inbasekaran, J. O'Brien, W. Wu, *Adv. Mater.* **12**, 1737 (2000).
8. W. Wu, M. Inbasekaran, M. Hudack, D. Welsh, W. Yu, Y. Cheng, C. Wang, S. Kram, M. Tacey, M. Bernius, R. Fletcher, K. Kiszka, S. Munger, J. O'Brien, *Microelectron. J*, **35**, 343 (2004).
9. C. R. Towns, I. Grizzi, M. Roberts, A. Wehrum, *J. Lumin.*, **122**, 976 (2007).
10. M. A. Baldo, S. Lamansky, P. E. Burrows, M. E. Thompson, S. R. Forrest, *Appl. Phys. Lett.* 75, **4** (1999).
11. X. H. Yang, D. Neher, D. Hertel, T. K. Daubler, *Adv. Mater.* **16**, 161 (2004).
12. D. Neher, *Macromol. Rapid Commun.*, **22**, 1365 (2001).
13. I. H. Campbell D. L. Smith, in *Solid State Physics* Vol. 55, Academic Press (2001).
14. S. R. Tseng, Y. S. Chen, H. F. Meng, H. C. Lai, C. H. Yeh, S. F. Horng, H. H. Liao, *Synth. Met.* **159**, 137 (2009).
15. S. R. Tseng, H. F. Meng, C. H. Ye, H. C. Lai, S. F. Horng, H. H. Liao, C. S. Hu, L. C. Lin, *Synth. Met.*, **158**, 130 (2008).
16. T. W. Lee, M. G. Kim, S. Y. Kim, S. H. Park, O. Kwon, T. Noh, T. S. Oh, *Appl. Phys. Lett.*, **89**, 123505 (2006).
17. S. R. Tseng, S. C. Lin, H. F. Meng, H. H. Liao, C. H. Ye, H. C. Lai, S. F. Horng, *Appl. Phys. Lett.* **88**, 163501 (2006).
18. J. Huang, G. Li, E. Wu, Q. Xu, Y. Yang, *Adv. Mater.* **18**, 114 (2006).
19. C. D. Muller, A. Falcou, N. Reckefuss, M. Rojahn, V. Wiederhirn, P. Rudati, H. Frohne, O. Nuyken, H. Becker, K. Meerholz, *Nature*, **421**, 829 (2003).
20. S. R. Tseng, S. Y. Li, H. F. Meng, Y. H. Yu, J. M. Yang, H. H. Liao, S. F. Horng, *J. Appl. Phys* **101**, 084510 (2007).

21. S. R. Tseng, S. Y. Li, H. F. Meng Y. H. Yu, C. M. Yang, H. H. Liao, S. F. Horng, C. S. Hsu, *Org. Electron.*, **9**, 279 (2008).
22. S. R. Tseng, H. F. Meng, K. C. Lee, S. F. Horng, *Appl. Phys. Lett.* **93**, 153308 (2008).
23. J. D. You, S. R. Tseng H. F. Meng, F. W. Yen, I. F. Lin, S. F. Horng, *Org Electron*, **10**, 1610 (2009).
24. C. G. Zhen, Z. K. Chen, Q. D. Liu, Y. F. Dai, R. Y. C. Shin, S. Y. Chang, J. Kieffer, *Adv. Mater.* **21**, 2425 (2009).
25. M. Mathai, V. E. Choong, W. Choulis, B. Krummacher, F. So, *Appl. Phys. Lett.*, **88**, 243512 (2006).
26. F. Huang, P. I. Shih, C. F. Shu, Y. Chi, A. K. Y. Jen, *Adv. Mater.* **21**, 361. (2009)
27. H. Wu, G. Zhou, K. Zou, C. L. Ho, W. Y. Wong, W. Yang, J. Peng, Y. Cao, *Adv. Mater.* **21**, 4181 (2009).
28. S. A. Choulis, V. E. Choong, M. K. Mathai, F. So, *Appl. Phys. Lett.*, **87**, 113503 (2005).
29. X. H. Yang, D. C. Muller, D. Neher, K. Meerholz, *Adv. Mater.* **18**, 948 (2006).
30. H. F. Meng (unpublished results).
31. S. C. Lo, N. A. H. Male, J. P. J. Markham, S. W. Magennis, P. L. Burn, O. V. Salata, I. D. W. Samuel, *Adv. Mater*, **14**, 975 (2002).
32. L. C. Ko, T. Y. Liu, C. Y. Chen, C. L. Yeh, S. R. Tseng, Y. C. Chao, H. F. Meng, S. C. Lo, P. L. Burn, S. F. Horng, *Org Electron.*, **11**, 1005 (2010).
33. M. R. Tseng, H. F. Meng (unpublished results).
34. F. I. Wu, H. J. Su, C. F. Shu, L. Luo, W. G. Diau, C. H. Cheng, J. P. Duan, G. H. Lee, J. *Mater. Chem.* **15**, 1035 (2005)..
35. C. Huang, C. G. Zhen, S. P. Su, Z. K. Chen, X. Liu, D. C. Zou, Y. R. Shi, K. P. Loh, *J. Organomet. Chem.*, **694**, 1317 (2009).
36. T. Tsuzuki, S. Tokito, *Appl. Phys. Express*, **1**, 021805 (2009).
37. L. Wang, B. Liang, F. Huang, J. Peng, Y. Cao, *Appl. Phys. Lett.*, **89**, 151115 (2006).
38. W. Song, J. S. Park, C. H. Kim, M. J. Im, J. S. Kim, Y. S. Gal, J. W. Kang, J. W. Lee, S. H. Jun, *Org. Electron.*, **10**, 1412 (2009).
39. Yen and Meng (unpublished results).

Chapter 3

Polymer Field-Effect Transistors

Yong-Young Noh,[a,*] Jui-Fen Chang,[b] Xiaoyang Cheng,[b] and Kang-Jun Baeg[c]

[a] *Department of Chemical Engineering, Hanbat National University, Daejeon 305-719, Republic of Korea*
[b] *Cavendish Laboratory, University of Cambridge, Cambridge, United Kingdom*
[c] *Heeger Center for Advanced Materials, Department of Materials Science and Engineering, Gwangju Institute of Science and Technology (GIST), Gwangju 500-712, Republic of Korea*
yynoh@hanbat.ac.kr

This chapter gives an overview of the theoretical background of polymer field-effect transistors (PFETs). It starts with the description of the molecular and electronic structures and optical transitions in conjugated polymers which are used as an active layer of PFETs. The second part of the chapter focuses on the working principles of PFETs including metal-insulator–semiconductor capacitors. Finally, several charge transport models developed for polymer and organic field-effect transistors and a few recent progress are discussed.

3.1 Introduction

Following the initial studies of electrical conductivity in organic crystals [1, 2], organic semiconductors have become an attractive

Polymer Electronics
Edited by Hsin-Fei Meng

ISBN 978-981-4267-84-7 (Hardcover), 978-981-4364-04-1 (eBook)
www.panstanford.com

class of active elements for electronic applications. Particular interest in conjugated polymers emerged after the 1976 discovery of the electrically conducting polymer, doped polyacetylene [3], which was recognized in 2000 with the Nobel Prize in Chemistry. Over the last 20 years much effort has been made in both academia and industry to investigate π-conjugated polymers for their semiconducting properties. In 1989, the discovery of electroluminescence in conjugated polymers was a major breakthrough [4], bringing commercial applications of conjugated polymers within reach for the first time in (opto)electronic devices, such as light-emitting diodes (LEDs) [5], photovoltaic cells (PVs) [6], and field-effect transistors (FETs) [7]. The potential of these polymer-based devices lies in the possibility to combine the advantages of solution processability, chemical tunability, and material strength of polymers with the typical properties of plastics to realize low-cost, large-area electronic devices on flexible substrates by solution deposition and direct-write printing techniques [8]. The recent progress on polymer synthesis, realization of *n*-type and ambipolar organic semiconductors, as well as scientific understanding of the device physics of these materials has flourished the development in the fields of plastic electronics and photonics [9].

In all of the applications, efficiency of charge transport within the semiconducting polymers plays a key role. Current solution deposition methods for polymer films produce complex microstructures with the inevitable presence of disorder. This results in complex physical systems in which charge transport properties depend critically on a variety of factors, such as the degree of interchain interaction, ordering of the chains, as well as on the density of chemical and structural defects [10–12]. To achieve high charge-carrier mobilities in solution-processible polymers, particularly important for use in thin film transistors (TFTs), two different approaches have emerged. The first approach is based on design of the polymers with microcrystalline or liquid-crystalline order through self-organization. The microcrystalline polymers, such poly(3-hexylthiophene) (P3HT) [13], poly(5,5'-bis(3-alkyl-2-thienyl)-2,2'-bithiophene) (PQT-12) [14], and poly(2,5-bis(3-alkylthiophene-2-yl)thieno[3,2-b]thiophene) (pBTTT) [15], have been demonstrated to exhibit high field-effect mobilities above 0.1 cm^2/Vs, approaching

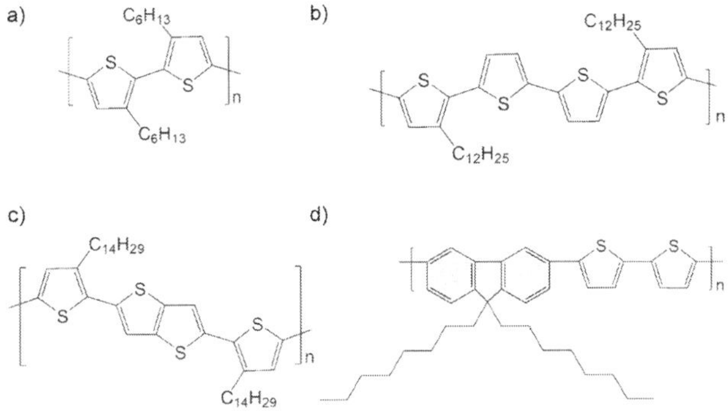

Figure 3.1 Chemical structure of (a) P3HT, (b) PQT-12, (c) pBTTT, and (d) F8T2.

the level of amorphous silicon. The liquid-crystalline polymers such as poly(9,9-dioctyl-fluorene-co-bithiophene) (F8T2) also have shown relatively high mobilities of 10^{-2} cm^2/Vs [16]. The second approach is to produce a completely amorphous microstructure by using specific interactions with a templating substrate. In this case the carriers may transport through a uniform path with a minimum degree of site-energy fluctuations. Some devices with field-effect mobilities of 10^{-3} to 10^{-2} cm^2/Vs and good environmental stability have been demonstrated with this approach [17]. The molecular structure of P3HT, PQT-12, pBTTT, and F8T2 are described in Fig. 3.1.

Parallel to the above work on conjugated polymers, research investigating thin film devices based on small conjugated molecules [18] and oligomers [19] has been performed. These materials can be deposited from vacuum sublimation and usually show higher carrier mobilities as a result of the high degree of molecular order. However, the fabrication costs are higher and the semiconducting films are not as robust as polymers, thus somewhat reducing their commercial appeal. Some recent efforts in this field have, therefore, focused on developing the solution-processible small molecules such as pentacene [20] and rubrene [21].

P3HT has self-organizing properties, and can form microcrystalline structures comprising two-dimensional conjugated lamellae with strong π–π interchain interactions in one direction and layers of insulating alkyl side chains in the other [7, 13]. In the past decade numerous studies have been made to optimize the mobilities and clarify the charge transport mechanisms in microcrystalline P3HT. It is well understood that the field-effect mobility of P3HT TFTs depends critically on the orientation of the conjugated lamellae and π–π stacking with respect to the substrate. This can be controlled by varying the degree of regioregularity [7], through dielectric modifications [22], by changing solvents [13], and using different deposition conditions [23]. The highest field-effect mobility of 0.1–0.3 cm^2/Vs has been obtained through optimized crystallinity of interchain packing [22]. In addition to crystallinity, recent research based on morphological studies pointed out that molecular weight is another important factor which affects mobilities in a regime with mobilities less than 10^{-2} cm^2/Vs [24, 25]. Two different mechanisms relating to either the degree of interconnectivity of the polymer network, or to the backbone polymer conformation were proposed to explain the observed molecular weight dependence of mobility.

Although previous work sheds light on the possibilities of pursuing even higher mobilities of polythiophenes by controlling structural anisotropy and molecular chain arrangement with strong interchain interactions, to date an understanding of the factors that limit the field-effect mobility in P3HT to values around 0.1 cm^2/Vs is still lacking, particularly in terms of morphology and molecular weight analysis. This fact may stem from the difficulty to modify the complex microstructure of solution-deposited P3HT formed in the solidification process as a result of the interplay between crystallization conditions, molecular weights, and interaction of polymers and solvents. Therefore, more detailed studies are required to understand how effects such as chemical modification of molecular weight or physical manipulation of microstructures eventually influence the charge transport in P3HT TFTs.

3.2 Introduction to Conjugated Polymers

Conjugated polymers are carbon-based, long-chain molecules with semiconducting electrical properties, which have attracted attention from many research groups and industry over the past 20 years. The critical feature that differentiates conjugated polymers from other insulating polymers is the presence of alternating single and double bonds, which are formed by the overlap of electronic orbitals from nearby atoms. The high photoluminescence/electroluminescence yield and strong absorption in the visible region of conjugated polymers make them highly popular for applications of LEDs and PVs [5, 6]. In addition, the promising charge carrier mobilities of conjugated polymers allow them to be utilized in FETs [7]. Although the semiconducting properties of conjugated polymers cannot currently challenge the high carrier mobilities of inorganic semiconductors, they have several advantages such as solution processable to fabricate low-cost, large-area electronic circuits on flexible substrates. Since their electronic properties strongly depend on their chemical structures and ordering, they are open to be specifically synthesized and optimized for the requirements of research and manufacturing.

3.2.1 *Electronic Structure*

The ground state of one carbon atom takes the electronic configuration $1s^2 2s^2 2p^2$. However, in the presence of the potential of other atoms, the nearly degenerate 2s and 2p states can form hybrid orbitals in turn to overlap more strongly with neighboring atoms. For example, in diamond or methane, the 2s and three 2p orbitals can be hybridized into four degenerate sp^3 hybrid orbitals arranged tetrahedrally around the central carbon atom. These orbitals form σ-bonds by overlapping with the electronic wave functions from neighboring atoms. In the case of conjugated polymers such as ethene ($H_2C{=}CH_2$), only the 2s and p_x and p_y orbitals are hybridized, forming three coplanar sp^2 orbitals directed about 120° apart from each other. The p_z orbitals are unaltered and oriented perpendicular to the plane of hybridization. The sp^2 orbitals form σ-bonds with neighboring atoms, while the overlap of p_z orbitals between adjacent sp^2 hybridized carbons forms π bonds. These π bonds establish a more delocalized electronic density than σ-bonds, and produce a filled bonding π orbital occupied by an electron from each atomic p_z orbital and an unfilled anti-bonding π^* orbital. As a result, in conjugated polymers the electronic properties are primarily based

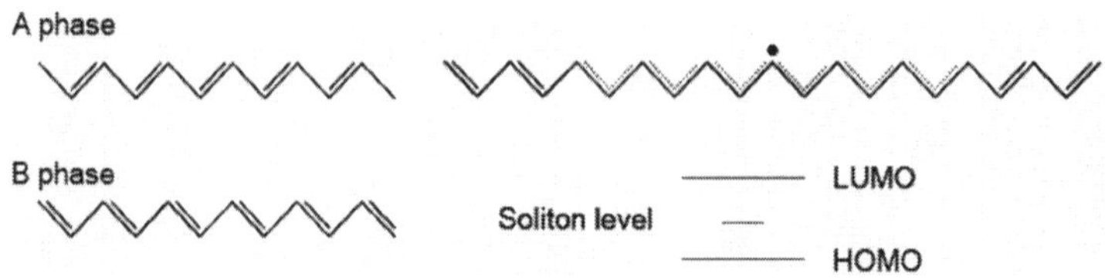

Figure 3.2 (Left) Bond conformation in polyacetylene with two degenerate bond alternations at ground states, A and B phases, of *trans*-polyacetylene. (Right) Physical and electronic structure of a soliton in acetylene. The mid-gap soliton energy level is formed.

on the electrons in π bond orbitals, and the molecular structure is based on the σ-bond orbitals.

If the chain is extended from two sp^2 hybridized carbons to an infinite number of sp^2 hybridized carbons which are equally spaced, a band structure will be formed. In principle, one might expect metallic behavior for such a half-filled band. However, in 1955 Peierls [26] showed that dimerization of the bond length by alternating single and double carbon bonds will result in the formation of an energy gap and create a filled π band and an unfilled π^* band. This leads to the semiconducting electronic properties of conjugated carbon chains, of which the simplest structure is in polyacetylene (Fig. 3.2). Polyacetylene has two degenerate ground states in which the bond alternation is opposite (termed A phase and B phase in Fig. 3.2). When these two states exist on the same chain, a bond alternation defect is formed in the middle of the two phases. In 1979, Su, Schriefer, and Heeger (the SSH model) [27] modeled this defect by considering the creation and annihilation of π electrons coupled to distortions in an infinite polyacetylene chain. They predicted that the defect as a soliton can move freely along the chain without dissipation of energy, and a half-occupied soliton energy level is formed in the middle of energy gap, i.e., the π electrons can move freely along the backbone in the ground state.

The work on the electronic structure of a benzene molecule is more relevant towards understanding conjugated polymers commonly used for semiconductor devices. For one benzene molecule, the orbitals of the carbon atoms around the benzene ring overlap

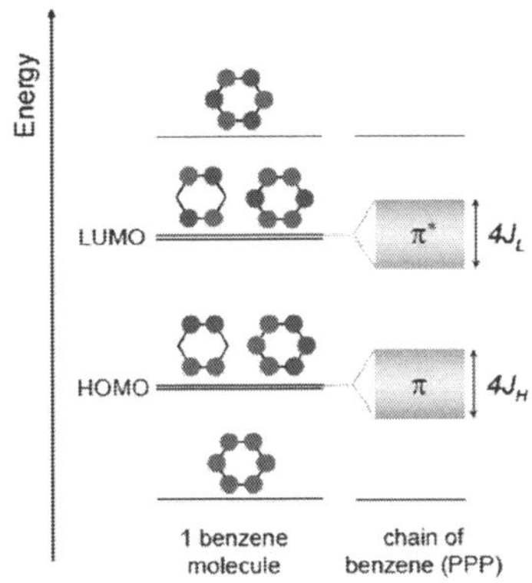

Figure 3.3 The π electron energy levels in a benzene molecule and a polymer chain of benzene molecules, calculated using the Hückel approximation.

with the π electrons interacting, forming discreet energy levels of bonding and antibonding orbitals. Within the Hückel approximation [28], a benzene molecule in the ground state has six different configurations in four different energy levels, two of which have a degeneracy of two (Fig. 3.3). The highest occupied molecular orbital (HOMO) is at the lower degenerate state with a bonding π orbital, while the lowest unoccupied molecular orbital (LUMO) is at the higher degenerate state with an anti-bonding π^* orbital. When an isolated benzene molecule is polymerized, as the case in poly(paraphenylene) (PPP), the HOMO and LUMO levels broaden into band states analogous to the valence band and conduction band of an inorganic semiconductor. The bandwidth of these quasibands is four times the transfer integral, J, between neighboring molecular states. The energy gap is in the visible range and depends on the length of the chain. Just as the simple particle in a box model, increasing the number of benzene rings on the chain reduces the energy gap between the HOMO and LUMO. However, in realistic conjugated polymer systems where the chains are long enough to have conformational defects, the chain is composed of manifold segments of conjugation lengths (the length of defect-free oligomer). These each have their own energy gap between HOMO and LUMO, leading to the broadening of π–π^* optical transitions.

3.2.2 *Neutral Excited States and Polarons in Conjugated Polymers*

Through the absorption of a photon by the conjugated polymer, one electron can be lifted from the filled HOMO level to the empty LUMO level. The resulting excited state is called an exciton and can be represented as a bound electron–hole pair. The exciton is associated with a lattice distortion and creates new energy levels within the HOMO–LUMO bandgap (Fig. 3.4a,b). The electronic transitions are generally completed before the molecule arriving the vibrational ground state of the new electronic state (Franck–Condon principle), because they are much faster than the nuclear motions. Therefore the electron can be excited to upper vibrational states, then relaxes radiationless to the lowest vibrational state, and finally returns to the ground state with emission of light (photoluminescence) or a radiationless channel of decay. The difference of the wavelength between the emitted light and the absorbed light is called the Stokes

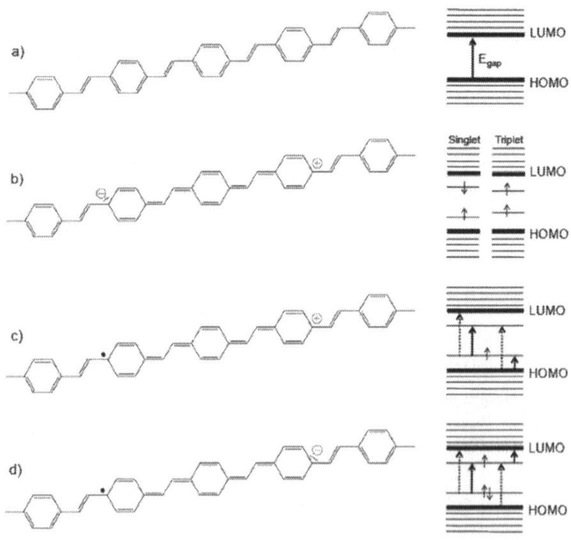

Figure 3.4 Illustration of bond alternation from (a) benzenoid in PPV to (b) quinoid in a neutral exciton, and in (c) positively or (d) negatively charged polaron including energy levels and allowed (solid arrows)/forbidden (dashed arrows) optical transitions.

shift, and the energy transfer from higher energy levels to lower energy levels of conjugated polymer may additionally contribute to the shift between absorption and emission [29]. There are two possible excited electronic states, named as the singlet and triplet states where the spin orientation of the bound electron and hole are antiparallel and parallel, respectively (Fig. 3.4b). Due to spin conservation, only singlets are formed during optical excitation and decay by nonradiative and radiative decay, and then these may undergo intersystem crossing to form triplets, although this process will be very slow. The fluorescence represents the radiative decay of singlets while the phosphorescence represents the radiative decay of triplets. Due to the forbidden transition between the triplet state and the ground state, the lifetime of triplets is usually longer than singlets [29].

During the operation of a FET, charge carriers are accumulated and transported within the semiconducting film layer, and the presence of either a positive or negative charge may cause a local structural relaxation of the polymer chains close to the charge because of the electron–phonon coupling. This kind of quasiparticle combining charge and lattice distortion is called a polaron for the many conjugated polymers with a nondegenerate ground state. For example, the presence of a positive charge could change the local configuration of a PPV chain from benzenoid to quinoid (Fig. 3.4c). The creation of a polaron state also results in the creation of two new energy levels within the bandgap as shown in Fig. 3.4c,d. In the electron polaron state one electron occupies the high-energy level while in a hole polaron state a hole exists in the low-energy level. These states will allow new electronic transitions with observed absorption features. However, only some of these transitions are symmetry allowed [30]. Figure 3.4 illustrates the HOMO/LUMO levels and polaron levels of a conjugated polymer with a positive or negative charge and the allowed or forbidden transitions. Further adding another charge to the polymer chain would lead to the formation of a spinless bipolaron with an even stronger lattice distortion and energy states deeper in the bandgap. These new optical transitions can be observed through charge modulation spectroscopy to study charge accumulation in organic field-effect transistors (OFETs) [31].

3.3 Polymer Field-Effect Transistor

3.3.1 *Metal–Insulator–Semiconductor Capacitor*

A typical metal-insulator–semiconductor (MIS) capacitor is constructed with gate electrodes (i.e., heavily doped Si used in the present work), a dielectric insulator, and a semiconductor, as shown in Fig. 3.5. For an ideal MIS capacitor, the bands are flat with no external voltage applied. This is the case when the modified work function $e\phi_m$ at the metal–insulator interface is equivalent to the modified work function $e\phi_s$ at the insulator–semiconductor interface, i.e.,

$$e\phi_{\mathrm{m}} = e\phi_{\mathrm{s}} = e\chi + \frac{E_{\mathrm{g}}}{2} + e\phi_{\mathrm{F}}. \tag{3.1}$$

Here χ is the potential difference between conduction band of semiconductor and that of insulator, E_g the semiconductor bandgap, ϕ_F the position of the Fermi level, E_F, below the intrinsic level E_i (assumed to be located close to the midgap) for the semiconductor. For a nonideal capacitor, the modified work functions differ at two interfaces, leading to a band bending at the insulator–semiconductor interface to accommodate the work function difference at equilibrium.

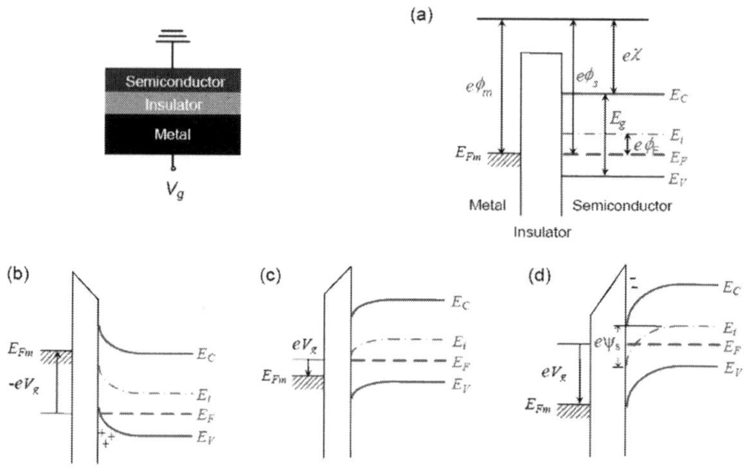

Figure 3.5 (Upper left) Schematic view of an MIS capacitor. Energy band diagram for ideal MIS diodes at equilibrium (a), and when $V_g \neq 0$ for (b) accumulation (c) depletion (d) inversion regimes.

In such a case, a potential $V_{FB} = \phi_m - \phi_s$ has to be applied to the metal to obtain the flat band condition as shown in Fig. 3.5a.

Three different situations may occur when a gate-bias V_g is applied to modulate the surface charge density at the insulator–semiconductor interface. Figure 3.5b–d illustrates a conventional band diagram of *p*-type inorganic semiconductors for the three regimes. Although the band structure of conjugated polymers is not the same as inorganic semiconductors, the HOMO and LUMO levels in conjugated polymers may be considered as band extremes. When applying a negative bias V_g to the gate electrode, the Fermi level of the metal is raised by eV_g so that holes are attracted to the semiconductor–insulator interface and therefore form an accumulation layer. This bias condition is the so-called accumulation regime (Fig. 3.5b). On the other hand, depletion regime occurs as a moderate positive bias is applied to the gate such that the majority carriers are depleted from the semiconductor–insulator interface and a space charge region is established (Fig. 3.5c). If applying a larger positive gate bias such that the bands bend downward and cross the Fermi level, minority carriers will be induced at the semiconductor–insulator interface and change the active layer from *p*-type to *n*-type, which is the so-called inversion regime (Fig. 3.5d). Inorganic MISFETs (metal–insulator–semiconductor field-effect transistors) usually operate in the inversion and depletion regimes. However, most of organic TFTs operate in the accumulation and depletion regimes, where the conductivity of the channel is often *p*-type since electrons tend to be trapped at the semiconductor/SiO_2 interface and are difficult to inject from standard electrodes (e.g., Au). Recently, several organic semiconductors with high electron affinity have been observed to exhibit *n*-type conductivity (see the review by Sirringhaus [9] and Dimitrakopoulos *et al.* [32]). Chua *et al.* further showed that the electron conductivity is a generic feature of most organic semiconductors by using the proper dielectric to avoid the trapping of electrons at the interface [33].

3.3.2 *Field-Effect Transistor*

Most OFETs are fabricated according to the geometry of the TFT. The basis scheme of a TFT is by adding two electrode contacts, the source and drain, on the semiconductor side of a MIS diode,

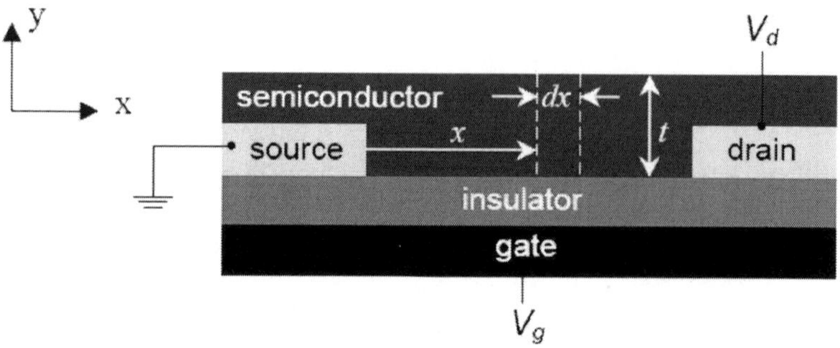

Figure 3.6 Schematic view of an organic field-effect transistor (OFET).

as depicted in Fig. 3.6. Two independent voltages drive a FET: the voltage applied to the gate across the insulator modulates the charge carrier density at the insulator–semiconductor interface, and the bias applied between source and drain drives the current induced by the gate bias.

The usual way to calculate the current–voltage (I–V) characteristics of a FET consists of estimating the elemental resistance dR of an elemental segment dx of the conducting channel [34]

$$dR = \frac{dx}{W\mu(x)\,|Q(x)|} \tag{3.2}$$

where W is the channel width, μ is the carrier mobility, and $Q(x)$ is the surface charge at x. In a TFT, $Q(x)$ is a summation of the bulk charge, Q_0, and the charge in the accumulation layer, $Q_s(x)$. Q_0 is a constant given by

$$Q_0 = \pm e n_0 t, \tag{3.3}$$

where n_0 is the density of free carriers at equilibrium, and t is the thickness of the semiconductor. The sign of the left-hand side depends on the majority carriers. $Q_s(x)$ is given by

$$Q_s(x) = -C_i[V_g - V_{FB} - V(x)]. \tag{3.4}$$

Here V_{FB} is the flat band voltage which accounts for the work function difference between semiconductor and the gate metal. $V(x)$ accounts for the drain voltage drop in the channel. In the linear regime where $V_d < V_g, V(x)$ gradually increases from 0 to V_d when moving from the source and the drain. Note that when dealing with an accumulation layer, $Q_s(x)$ and Q_0 have identical sign. Therefore, using Eqs. (3.24) and (3.25) and Ohm's law, Eq. (3.23) can be rewritten as

$$dV = I_d^{lin} dR = \frac{I_d^{lin} dx}{W\mu(x) \left| C_i(V_g - V_{FB} - V(x)) + en_0 t \right|}. \quad (3.5)$$

Integrating from source ($x = 0$, $V = 0$) to drain ($x = L$, $V = V_d$) we obtain

$$I_d^{lin} \int_0^L dx = I_d^{lin} L = W \int_0^{L_d} \mu(x)[C_i(V_g - V_{FB} - V)] + en_0 t] dV. \quad (3.6)$$

Assuming that the mobility is constant along the channel, integration of Eq. (3.27) then yields

$$I_d^{lin} = \frac{WC_i\mu}{L}\left((V_g - V_0)V_d^{-}\frac{V_d^2}{2}\right). \quad (3.7)$$

Here $V_0 = \pm en_0 t/C_i + V_{FB}$ is defined with a similar concept of "threshold voltage" in inorganic transistors, which accounts for the nonzero drain current flow in the device at zero gate bias and its sign is inversed to V_g The linear mobility μ_{lin} can be therefore estimated from the transconductance obtained by differentiating the Eq. (37) with respect to the gate voltage,

$$g_m^{lin} \equiv \left.\frac{\partial I_d^{lin}}{\partial V_g}\right|_{V_d=\text{const.}} = \frac{W}{L}\mu^{lin} C_i V_d. \quad (3.8)$$

When $V_d > V_g$ the accumulation layer extending from the source is pinched off at a point where $V(x) = V_g$, beyond which it is replaced by a depletion layer. In this depletion region, there is no free charge at the insulator–semiconductor interface and the effective thickness of the bulk semiconductor film is reduced. The drain current in this region is given by

$$I_d dx = W\mu(x) en_0[t - W_d(x)]dV, \quad (3.9)$$

where W_d is the width of the depletion layer

$$W_d(V) = \frac{\varepsilon_s}{C_i}\left[\sqrt{1 + \frac{2C_i^2(V(x) - V_g)}{eN\varepsilon_s}} - 1\right]. \quad (3.10)$$

Here ε_s is the permittivity of the semiconductor and N is the concentration of dopants. Note that V_{FB} is neglected in Eq. (3.10).

The saturation current is hence given by the sum of integrals in the accumulation and depletion regions [35],

$$I_d^{Sat} L = WC_i \int_0^{V_s} \mu(x)(V_g - V_0 - V)dV + Wen_0 \int_{V_s}^{V_d} \mu(x)[t - W_d(x)]dV. \quad (3.11)$$

Assume that the second term is integrated from $W_d(V_g) = 0$ to $W_d(V_d) = t$ and the mobility is constant along the channel, then the integration in Eq. (3.11) is obtained as

$$I_d^{Sat} = \frac{W}{L}\mu \left[C_i \left(\frac{V_g^2}{2} - V_0 V_g \right) + \frac{e^2 n_0 N\, t^3}{\varepsilon_s\, 6} \left(1 + 3\frac{C_s}{C_i} \right) \right], \quad (3.12)$$

where $C_s = \varepsilon_s/t$ is the dielectric capacitance of the semiconductor. If we define a pinchoff voltage V_p as the gate bias applied to completely deplete the semiconducting layer, from Eq. (3.10) we could obtain [34]

$$V_P \approx \pm \frac{eNt}{C_i}. \quad (3.13)$$

With V_p the saturation current in Eq. (3.12) can be simplified as

$$I_d^{Sat} = \frac{W}{2L}\mu C_i (V_g^2 - 2V_0 V_g + V_0 V_p). \quad (3.14)$$

Assuming $n_0 = N$ then yields $V_0 \approx V_p$, and the equation can be further simplified to

$$I_d^{Sat} = \frac{W}{2L}\mu C_i (V_s - V_0)^2. \quad (3.15)$$

This equation is frequently used to estimate the saturation mobility of OFETs through a plot of $\sqrt{I_d^{sat}}$ as a function of V_g. By fitting this curve with a straight line, the intercept on the V_g axis determines V_0, and the slope of the fitted line would be proportional to the mobility. However, it is important to note that the validity of Eqs. (3.7) and (3.15) rests on the assumption of a constant mobility, or at least to a first approximation of a slowly varying gate voltage dependent mobility. In some highly crystalline OFETs a linear dependence of $\sqrt{I_d^{sat}}$ vs. V_g is usually observed, indicating that the saturation mobility is approximately gate voltage independent and thus can be estimated from Eq. (3.15). By contrast, most disordered

OFETs exhibit a superlinear increase of $\sqrt{I_d^{sat}}$ wit V_g, indicative of a strong gate voltage dependent mobility. In such a case, Eq. (3.15) is inapplicable to extract the saturation mobility and also the value of V_0 might yield an unreasonably high free carrier density [34]. Alternatively, more accurate gate voltage dependent mobilities can be achieved by estimating the mobility in the linear regime through Eq. (3.8).

The origin of gate voltage dependent mobility in organic materials stems from a high density of localized states due to their prevalent structural disorder. This can be rationalized with the multiple trapping and release (MTR) model [36], which assumes a large concentration of localized states lying below the conduction band. At low gate voltage, most of the charges are trapped by these states, thus leading to a low mobility. As the gate voltages increases more traps are filled, and the Fermi level moves closer to the conduction bands leading to an increased carrier concentration in this band and subsequent rise of the mobility. Eventually, all the trap states are filled and any additional charge can move freely in the conduction band with mobility independent of the gate voltage.

Moreover, thermally activated mobility is observed in general OFETs. This is unlike the band-like transition in inorganic FETs where the transport is primarily limited by electron–phonon interactions and so the mobility increases with decreasing temperature. In OFETs the mobility decreases and becomes more superlinear with gate voltage as temperature decreases. Again, Eq. (3.15) becomes inapplicable for estimation of the saturation mobility at low temperature. Instead, most published work has studied the temperature dependence of linear mobility through Eq. (3.8), or used a different description of "effective mobility" [37]. The thermally activated mobility can be attributed to intrinsic charge transport mechanisms in organic materials. Charges in conjugated polymers may relax onto molecules forming self-localized polarons, and the transport of these small polarons occurs via thermally activated hopping. On the other hand, the MTR model suggests that the charges trapped by the localized defects can eventually be thermally released to the conduction band. This release mechanism together with the small polaron formation results in a general observation of the thermally activated mobility for OFETs. The temperature and gate voltage dependence of mobility in organic semiconductors is closely related to the distributions of localized states in the gap, which varies with structural order associated with different processing conditions [38].

3.3.3 *Contact Resistance*

Previous discussions based on current–voltage equations assume no potential drop at the source–drain electrodes for either charge injection or charge extraction, however in practice this is not a realistic FET working model due to the existing contact resistance, which represents a potential drop at the source–drain electrodes preventing the full source–drain bias being applied across the channel. Usually the contact resistance values in organic FETs are on the order of 10 kΩ to 100 MΩ [39] which is considerably larger than those values in inorganic transistors. The reason is that in organic transistors the electrodes are not easily optimized by conventional processes such as selective doping of the semiconductor in silicon MISFETs. Large contact resistance can severely suppress the transistor characteristics if it becomes comparable to, or even exceeds the channel resistance, and results in an insufficient charge injection with a reduction in field-effect mobility [40, 41]. The most obvious feature of contact resistance is the superlinear increase of drain current at small drain voltages in the output characteristics. In this condition the total resistance of a FET with given channel length L can be expressed as

$$R_{\mathrm{total}}(L) = \left.\frac{\partial V_{\mathrm{d}}}{\partial I_{\mathrm{d}}}\right|_{V_{\mathrm{d}}\to 0} = R_{\mathrm{ch}}(L) + R_{\mathrm{c}}, \qquad (3.16)$$

where $R_{\mathrm{ch}}(L)$ represents the channel resistance and R_{c} represents the overall contact resistance. Thus we plot R_{total} as a function of channel length L; the intercept of the R_{total} at $L = 0$ gives the value of R_{c} and the slope of $R_{\mathrm{total}}(L)$ gives the mobility. This method is called the transmission line method (TLM) and it is the most common way for deducing contact resistance in the FETs due to its simplicity [42, 43]. However, the main drawback to this method is that it is only valid for Ohmic contacts and it cannot determine the contact resistance between source and drain electrodes. A more accurate way to measure the contact resistance in transistors is the gated four-point probe (gFPP) measurement [44–46]. A device for the gFPP measurement has two voltage probes in the transistor channel with a separation distance x. This is used to measure the potential at two points therefore it is possible to extrapolate the channel potential out to the contacts [46]. Using this method, the voltage drop across source and drain contacts can be measured and corresponding contact resistance values calculated. This provides an almost direct measurement of contact voltage to current and is valid for non-Ohmic contacts [46]. Another powerful technique to characterize contact resistance is the scanning Kelvin probe microscopy (SKPM), which can directly image the change of electrostatic potential inside an operating device and is therefore able to decouple the resistance of the contacts from that of the bulk [39].

There are generally two mechanisms that would contribute to the contact resistance: charge injection from the source–drain electrodes and charge transport though the bulk of semiconductor before they reach the accumulation layer or the electrodes [39]. For *n*-type transistors this applies to injection of electrons from the metal electrodes into the LUMO level of semiconductor while for *p*-type transistors injection of holes into the HOMO level of semiconductor. A charge injection energy barrier may arise from the energy difference between the work function of the injecting metal and the HOMO or LUMO level of the semiconductor, therefore the metal–semiconductor interface is usually treated as a Mott–Schottky barrier, where the barrier height is determined by the difference between the metal work function and the LUMO or HOMO levels of the semiconductor (Fig. 3.7a) [47]. There are several models which are proposed to describe the charge injection process for a Schottky contact, such as thermally assisted tunneling, thermally assisted injection into a distribution of localized states, and diffusion limited thermionic emission [48–50]. Using SKPM studies, Burgi *et al.* shows that contact resistance reflects Schottky barrier height, but also contains a dependence on bulk transport and mobility [39]. Additionally, they also demonstrated that comparing the work function of the injecting metal with the LUMO/HOMO levels of a semiconductor could help to determine whether charge carrier can

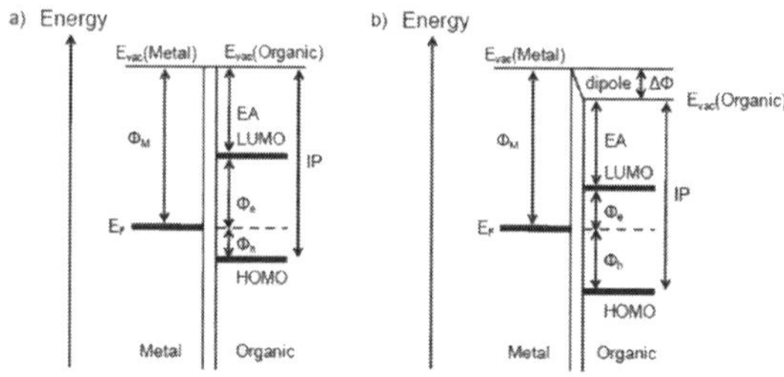

Figure 3.7 Energy level diagrams of interface between metal and semiconductor without a dipole layer a) and with a dipole layer b). E_F: Fermi energy, Φ_M: metal work-function, E_{vac}: vacuum level, EA: electron affinity, IP: ionization potential, Φ_e and Φ_h: electron and hole injection barrier respectively.

be injected and whether low or high contact resistance is expected. To reduce this Mott–Schottky barrier, either selective interfacial doping of semiconductor or inserting dipolar molecules with well aligned dipole moments at the metal interface have been widely employed, as shown in Fig. 3.7b [51–53].

The charge injection is not only related to the mismatch between the work function of metal electrodes and the HOMO/LUMO levels of semiconductors but also strongly depends on the geometry of the transistor. The physical nature of the semiconductor as well as the employed gate dielectric may require different transistor configurations which may show very different device performance. The most commonly used transistor configurations are the bottom-contact/bottom-gate (BC/BG, Fig. 3.8a), top-contact/bottom-gate (TC/BG, Fig. 3.8b), and bottom-contact/top-gate (BC/TG, Fig. 3.8c) structure.

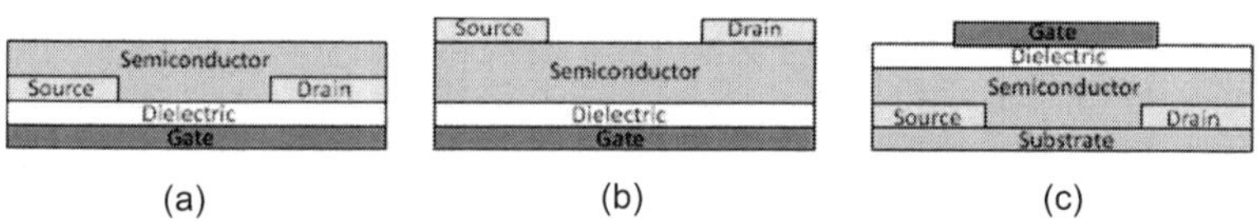

Figure 3.8 Common geometries of field-effect transistors: (a) bottom-contact/bottom-gate (BC/BG), (b) top-contact/bottom-gate (TC/BG), (c) bottom-contact/top-gate (BC/TG).

One of the major differences between the different transistor geometries arises from the position of the injecting electrodes in relation to the gate electrode. In the BC/BG configuration (coplanar structure), charges are directly injected from the vertical edge of the source electrode into the accumulation channel at the semiconductor/dielectric interface. However, in the other two configurations (TC/BG and BC/TG), because the injecting electrode and the accumulation channel are separated by the semiconductor, charges need to travel through a distance (normally several tens of nanometer) in the semiconductor before they reach the channel region. In the TC/BG (Fig. 3.8b) and BC/TG (Fig. 3.8c) configurations (staggered structure), charges are not only injected from the edge of the electrode but also from those parts of the electrode that overlap with the gate, which will contribute to the current flow depending on the distance from the edge. This leads to an increased injection current and current crowding [46].

Other differences in the transistor characteristics due to device geometries are common and they mainly originate from differences of the metal–semiconductor and semiconductor–dielectric interfaces, such as morphological effects of the semiconducting layer near contact region or introduction of trap states during the fabrication process [54]. Therefore, it is very important to select the suitable transistor configuration for the evaluation and comparison of organic semiconductors.

3.3.4 *Bias Stress*

The bias stress effect is a commonly observed problem during the operation of FETs, which appears as a threshold voltage shift together with a degradation of the output drain current. The bias stress can be caused by many reasons, such as mobile ions, the structural degradation of semiconducting layers or the trapping of charge carriers [55]. In the charge trapping mechanism, different trapping processes will lead to different variation in the transfer characteristics before and after stressing the device. If the charges are trapped in deep states, they will become immobile charges which shield the applied gate field, therefore a translation of the transfer characteristics in the direction of the gate bias will be observed.

This kind of bias stress effect usually occurs in organic and polymer FETs and has been widely discussed in detail [35, 56]. Alternatively if the charges are predominantly trapped in the shallow states, the transfer characteristics will display some hysteresis [57, 58].

In amorphous silicon transistors, the bias stress effect is generally explained by the creation of dangling bond states during the operation of the devices [59]. Contrary to that, several mechanisms of bias stress effects in organic transistors have already been proposed and discussed. In most *p*-type organic semiconductors a negative shift of the threshold voltage is found upon prolonged operation of the device in accumulation, which is generally attributed to charge trapping in the semiconductor and/or at the active interface. Brown *et al.* observed a negative threshold voltage shift under applied negative gate bias in pentacene transistors and showed that such effects are independent of the gate dielectric [35]. A similar phenomenon in F8T2 transistors has also been reported, and the authors found that the bias stress effect can be rapidly reversed under an illumination of light with energy equivalent to the bandgap of the conjugated polymer [60]. Furthermore, they demonstrated that the trapping rate of holes is proportional to the square of their concentration [61]. Based on these results, they proposed the bias stress effect is dominated by the formation of trapped bipolarons and used this mechanism to explain their observation. In polythiophene-based transistors, the bias stress effect is weak and transient because of a small bipolaron binding energy, while in F8T2 transistors the bipolarons are more stable and a large bias stress can be detected. Nevertheless Deng *et al.* had shown that no evidence of existing bipolarons in operating F8T2 transistors by using charge modulation spectroscopy [31]. Other studies of the bias stress effect in F8T2 and PQT-12 transistors were also performed by Ng *et al.* but they mainly focused on the effect of the gate dielectric surface on the kinetics of charge trapping [62]. They demonstrated the passivation of the oxide dielectric surface would reduce the bias stress effect by eliminating the deep trap states and minimizing the average time a charge resides in these traps. A further study of F8T2 transistors with different device geometry was reported by Richards *et al.*, where they indicated that contact resistance would also contribute to the bias stress effect [63]. It is clear from these experiments that the device configuration, choice of contacts and dielectrics play an important role in determining the device stability.

The bias stress effect is also intensively investigated in the small-molecule organic transistors [64–66]. In pentacene-based transistors, both positive and negative threshold voltage shifts have been measured depending on the applied gate bias [64]. The density functional calculations suggested that the hydrogen-induced defect states in pentacene could be responsible for the observation of bias stress effect [67]. Subsequently, Zilker *et al.* demonstrated that the direction of the threshold voltage shift can be reversed if the gate dielectric contains a large amount of water molecules [56]. In this condition the bias stress effect can be caused from the water-induced mobile ions in the gate dielectric, and the humidity is also turned to influence the operation stability of PQT-12 transistors.

3.4 Charge Transport Theory in Polymer FETS

3.4.1 *Disorder-Induced Hopping Models*

It is generally accepted that the charge transport in disordered organic solids occurs by hopping between localized states. Many of hopping theories originate from the Miller–Abrahams jump rate model [68], which was proposed for studying the conductivity through phonon-induced electron hopping in low-impurity inorganic semiconductors at low temperature. The localized states are shallow impurity levels. This model assumes that an uphill hop is characterized by a balance factor, $\exp(\Delta / k_B T)$, and downhill hop is temperature independent and not impeded by an energy-matching condition. The hopping rate between adjacent occupied site i and unoccupied site j separated in energy by E_{ij} and in distance by r_{ij} is given by

$$W_{ij} = \nu_0 \exp(-2\alpha r_{ij}) \times \begin{cases} \exp(-E_{ij}/k_B T) & E_{ij} > 0 \\ 1 & E_{ij} < 0 \end{cases}, \tag{3.17}$$

where ν_0 is a typical phonon frequency factor and α^{-1} is the localization radius of a charge carrier.

Mott extended the hopping mechanism with a variable range hopping (VRH) concept [69], where the charge carrier may either hop over a small distance with a high activation energy or hop over a long distance with a low activation energy in a constant density of localized states. Assuming that the wave functions are isotropic, the temperature dependence of mobility is derived as

$$\mu \propto \exp\left[-\left(\frac{T_0}{T}\right)^{1/n+1}\right], \tag{3.18}$$

where T_0 is a parameter inversely proportional to the density of states at the Fermi level and n is the dimensionality of the system ($n = 1, 2, 3$). For conjugated polymers where the wave functions are intrinsically anisotropic, it has been shown that the exponent of T_0/T lies between 1/2 and 1/4 [70]. The hopping through an exponential DOS studied by Monroe [71], however, showed that the characteristic hop is thermally activated since there are more available states at higher energies. There is no need to invoke a delocalized transport level. The mobility has Arrhenius-like temperature dependence

$$\mu \propto \exp\left(-\frac{E_a}{k_B T}\right), \tag{3.19}$$

where the activation energy, E_a, is weakly temperature dependent. In the framework of the exponential density of tail states, Vissenberg and Matters formulated the field-effect mobility in amorphous organic FETs by combining the VRH concept with percolation theory and a transistor model [72]. This model predicts a similar Arrhenius dependence, which is in agreement with the experimental data taken from pentacene and poly(thienylene vinylene).

In 1993 Bässler proposed a charge transport model for disordered organic systems [73]. He assumes that the carriers are not polaronic and the activation energy of the charge transport is entirely dominated by both energetic and positional disorder of the hopping sites. The energetic disorder, also called diagonal disorder, arises from fluctuation in site energy due to the random distribution of permanent dipoles and induced dipoles associated with the van der Waals forces. Fluctuations of intersite distances are otherwise termed as positional or off-diagonal disorder. The hopping rates between occupied site i and unoccupied site j are described by the Miller–Abrahams formalism (Eq. (3.17)). A Gaussian distribution of localized states was assumed

$$D(E) = \frac{N_t}{(2\pi\sigma)^2} \exp\left(-\frac{E^2}{2\sigma^2}\right) \quad , \tag{3.20}$$

where N_t is the total density of sites, σ is the width of the Gaussian density of state (DOS) and E is the energy relative to the center of the DOS. The Gaussian DOS shape is supported by the fact that the coupling between a charge carrier and a random distribution of permanent or induced dipoles has a Gaussian DOS function [74]. Monte Carlo simulations were used to solve the charge transport in Gaussian disorder model (GDM). The computational results showed that injected charges with arbitrary energy within a Gaussian DOS relax to a dynamic equilibrium energy $< \varepsilon_\infty >= -\sigma^2/k_B T$ below the center of the DOS and the energy required to transport in the transport level is -(5/9) $\sigma^2/k_B T$. The difference between the equilibrium energy and transport level leads to a non-Arrhenius temperature dependence of mobility $\mu \propto \exp[-(2\sigma/3k_B T)^2]$. Additionally, a Poole–Frenkel (PF)-type field dependence of mobility $\mu \propto \exp(\gamma E^{1/2})$ has been observed experimentally at high fields in many disordered organic materials [75–77]. In the high field range ($> 10^6$ V/cm) the field and temperature dependence of the mobility is often empirically characterized as

$$\mu = \mu_0 \exp\left[-\left(\frac{2\sigma}{3k_B T}\right)^2\right] \times \begin{cases} \exp\left[C\left(\left(\frac{\sigma}{k_B T}\right)^2 - \Sigma^2\right) E^{1/2}\right] & \text{for } \Sigma \geq 1.5 \\ \exp\left[C\left(\left(\frac{\sigma}{k_B T}\right)^2 - 2.25\right) E^{1/2}\right] & \text{for } \Sigma < 1.5 \end{cases} , \tag{3.21}$$

where μ_0 is the mobility at infinite absolute temperature and at zero field, C is an empirical constant, and Σ is the degree of positional disorder. Equation (3.21) predicts that the slopes of log μ vs. $E^{1/2}$ change sign at $\sigma/k_B T = \Sigma$ (for $\Sigma \geq 1.5$) or even in the absence of Σ. This complex field-dependent mobility relating to the degree of both diagonal and off-diagonal disorder can be realized in the following way. For systems without energetic and positional disorder ($\sigma = 0, \Sigma = 0$), the drift velocity of a carrier saturates with field at high fields because a carrier is only able to hop in the forward direction at a constant hopping rate. Therefore, mobility decreases as the electric field increases. For systems with sufficiently large energetic disorder and no positional disorder ($\sigma > 0, \Sigma = 0$), increasing a bias field reduces the activation energy for hops along the field direction, resulting in a mobility that increases with field. On the other hand, for systems with only positional disorder ($\sigma = 0, \Sigma > 0$), the mobility will decrease with field. This is because at low fields the carrier may choose an energetically easier path, even along loops against the field direction to arrive at a downstream site faster than without positional disorder. However, at high field the paths are constrained along the field direction, leading to decreased mobility. Energetic and positional disorder thus acts in an opposing way to influence the field-dependence of mobility.

In Bässler's GDM, site energies are distributed independently and without correlations occurring over any length scale. However, it has been observed that the charge–dipole interaction provides a significant contribution to the site energy [78]. Taking this into account, the site energy will vary slowly, and is correlated over a length scale which is greater than the intersite distance. Such a large scale spatial correlation is considered in the correlated disorder model (CDM) proposed by Novikov *et al.* [79]. Based on computer simulations of a 3D transport model, the empirical expression for the electric field and temperature dependence of the mobility becomes

$$\mu = \mu_0 \exp\left[-\left(\frac{3\sigma_\mathrm{d}}{5k_\mathrm{B}T}\right)^2 + 0.78\left(\left(\frac{\sigma_\mathrm{d}}{k_\mathrm{B}T}\right)^{3/2} - 2\right)\sqrt{\frac{eaE}{\sigma_\mathrm{d}}}\right], \quad (3.22)$$

where σ_d is the width of the dipolar energetic disorder, and a is the minimal charge dipole separation. The main difference between the two models is the range of fields over which PF behavior occurs. The GDM model shows PF behavior only over a very narrow range at high fields [8], while the CDM model shows that the PF field dependence is a universal feature extending to lower electric fields, which is associated more with long-range potential fluctuation rather than the kind of hopping rate. In recent years the CDM hopping model has been widely used to explain the temperature and field dependent mobilities in LED devices [81].

3.4.2 *Polaron Theory*

In a fully periodic system, the electronic properties can be described in terms of Bloch functions, energy bands, $E-k$ dispersion relations, and the electrons and holes can be regarded as free particle-like charge carriers. In this case, the mean free path of charge carriers is much greater than the lattice constant. For organic materials, the band transport has been reported in several high purity molecular single crystals such as naphthalene [82] and pentacene [83] at low temperature. In these molecular crystals the electron–phonon interaction is weak because the extended delocalization of electrons does not effectively disturb the crystal structure. However, adding local disorder in a periodic system will induce scattering centers and energy states in the forbidden bandgap, and decrease the mean free path of charge carries. The enhanced electron–lattice interaction causes a charge carrier to self-trap onto a single molecule, forming a composite quasiparticle consisting of a charge and the associated structural deformation. This quasiparticle is called a polaron. The polaron concept was originally conceived by Landau in 1933 [84], who showed that self-trapping would always occur in an ionic lattice. Fröhlich *et al.* [85] further advanced the theory of "large" polaron for ionic solids, where the polaron was assumed to be highly delocalized and with relatively small binding energy. However, the general case concerned in organic materials is a "small" polaron, where the dimensions of the deformation are of the order of an intermolecular spacing.

In the late 1950s, Holstein developed a molecular crystal model for small polaron formation and hopping in molecular solids [86, 87]. The model assumes that the Hamiltonian of the system is composed of three terms which account for lattice, electron, and electron-phonon coupling, respectively. The lattice component consists of a sum of harmonic oscillators with the same vibrational frequency. The dominant parameter of the electron component is the electron transfer integral, J, describing the motion of the electron through the lattice. The electron–phonon coupling is quantified by the polaron binding energy, E_{p}, which corresponds to the energy lowering due to the polarization and deformation of the lattice. The small polaron condition is realized when $2J << E_p$. In such a case the electronic term can be then treated as a perturbation in the 0th order Hamiltonian, which gives rise to two different transitions, involving either all of the quantum-numbers unchanged ("diagonal" transitions) or some of them changed by ± 1 ("nondiagonal" transitions) as the electron hops to neighboring sites. The diagonal transitions will give rise to Bloch-like polaron bands with a bandwidth, ΔE, given by the product of the electronic-overlap integral and a vibrational overlap integral. The nondiagonal transitions otherwise act as a scattering mechanism, which determines the life time of polaron band states and the mean free path of polaron transport. In the low-temperature regime, the diagonal transition dominates, while the probability of the nondiagonal transition is limited because of the reduced phonon scattering. The electron may diffuse via the narrow polaron band. The electron diffusivity, and hence the mobility, increases as temperature decreases. However, as temperature increases, the probability of the nondiagonal transitions increases exponentially and the bandwidth ΔE diminishes. Above one-half of the Debye temperature, the life time of the polaron band states becomes less than the time required for the polaron moving a distance equal to a lattice spacing. The band approach is no more applicable. The small polaron transport in the high-temperature regime (above the Debye temperature) is then dominated by thermally activated intersite hopping associated with the nondiagonal transitions.

In Holstein's model the successive hops are regarded as independent events. Here we use a semiclassical picture to explain the small polaron hopping at high temperature, as illustrated in Fig. 3.9. When a carrier sits on a molecule, the electron energy level is lowered with respect to that of an undeformed molecule (Fig. 3.9a). The hopping may occur when the lattices of two neighboring sites deform such that the two local electronic energy levels achieve a momentary energy degeneracy (coincidence event) (Fig. 3.9b). Such a coincidence is achieved by thermal deformation of the lattice. Then the hopping scheme is completed when the carrier equilibrates with the molecular vibrations at its newly occupied site (Fig. 3.9c). Quantitatively, the hopping rate is the product of (i) the probability of a coincidence event and (ii) the probability for a carrier to hop given a coincidence event, P. The steady-state mobility is then given

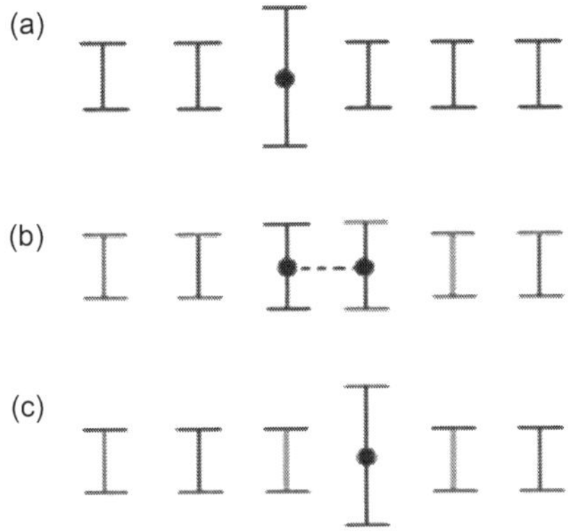

Figure 3.9 A schematic representation of the Holstein's small polaron hopping in the high-temperature regime. For a detailed description see text.

by the semiclassical formula

$$\mu = \frac{ea^2 v_0}{k_B T} P \exp\left(-\frac{E_a}{k_B T}\right), \tag{3.23}$$

where e is the absolute electron charge, a is the mean intermolecular distance, v_0 is the characteristic atomic vibrational frequency, $k_B T$ is the thermal energy and E_a is the activation energy which represents the minimum energy required to deform the molecules to form a coincidence event. Emin *et al.* divided the hopping processes into adiabatic and nonadiabatic cases [88]. When the electronic transfer integral J is so small that the electron cannot respond rapidly enough to the occurrence of a coincidence event, this is referred to as "nonadiabatic" hopping and

$$P = \frac{\sqrt{\pi} J^2}{\hbar v_0 \sqrt{4 E_a k_B T}} << 1. \tag{3.24}$$

The mobility in the nonadiabatic hopping then follows

$$\mu \propto \frac{1}{T^{3/2}} \exp\left(-\frac{E_a}{k_B T}\right). \tag{3.25}$$

If the values of J are sufficiently large that P in Eq. (3.24) exceeds unity, this is referred to as adiabatic hopping and P should be replaced by unity in Eq. (3.23). Therefore the temperature dependence of mobility in the case of adiabatic hopping is

$$\mu \propto \frac{1}{T} \exp\left(-\frac{E_a}{k_B T}\right). \tag{3.26}$$

3.4.3 *Multiple Trapping and Release Model*

The MTR model was developed by Le Comber and Spear for hydrogenated amorphous silicon (a-Si:H) [89]. The model assumes a high concentration of trap states below a well-defined transport level. The charges are trapped into the localized states with a high probability, and the transport can only occur when they are released to the transport level by thermal activation. Therefore, an Arrhenius temperature dependence of the mobility is predicted

$$\mu = \mu_0 \alpha \exp\left(-\frac{E_\mathrm{T}}{k_\mathrm{B}T}\right), \tag{3.27}$$

where μ_0 is the mobility above the transport level, α is the ratio between the effective density of states at the transport level to the density of traps, and E_T is the mean energy of the trap states [9]. The distributions of charges, and hence the mobility, are determined by distributions of traps, temperature, and voltage applied [36]. Despite the fact that the assumptions of a well-defined mobility edge, and a constant mobility above the transport level, should be too simple for disordered organic materials, the MTR model has been demonstrated to be able to reasonably capture the transport characteristics of high-mobility organic FETs [22, 38, 91].

3.5 RT Progress in Polymer Field-Effect Transistor

3.5.1 *Development of High-Performance n-Type Conjugated Polymers*

The logic elements are of crucial important beyond single transistor devices for the development of electronic circuits based on organic electronics. Inverters are the most fundamental circuits for logical functions and are widely applied in ring oscillators, which are components of various digital and analog circuits. In silicon-based technology, the inverters are complementary, which means they compromise both *p*-type and *n*-type transistor. Although it is possible to fabricate inverters with only one type of transistor (e.g., *p*-type) as many researchers already demonstrated, it is still advantageous to use complementary circuits for organic logic electronics [92–94]. Apart from the low power dissipation, the complementary circuits are easier to design and show more robust operation and better noise margins. In order to obtain organic complementary inverters, it is necessary to fabricate reliable organic *n*-type transistors which can be integrated with *p*-type transistors. Unfortunately, most of the high-performance organic transistors in the literature are *p*-type transistors [7, 15], and the need of building complementary circuits has led to research into developing *n*-type transistors over the past 10 years as well as organic complementary inverters and other complex circuits [95–99]. Although the majority of organic semiconductors seem to preferentially exhibit hole transport behavior (*p*-type), there is no clear reason why these materials do not show electron transport (*n*-type). Therefore it is essential to understand organic semiconductors and investigate the intrinsic or extrinsic reasons for the unlikely *n*-type behavior in organic FETs.

One of the most difficult challenges realizing *n*-type transistors is the electron injection from suitable metal electrode to the LUMO level of the organic semiconductors. In *p*-type transistors, the hole injection into the HOMO level is relatively easy to achieve by using a high work function metal such as gold. The HOMO level of most organic semiconductors including conjugated polymers is in the range of 4.8–5.5 eV, which aligns well with the work function of gold (4.7–5.5 eV) depending on different exposure conditions, and the LUMO levels for most organic semiconductors are often located much higher, at about 2.5–3.5 eV [100]. If gold contacts are employed in this case for injecting electrons, the observation of *n*-type transport is very challenging, as would be expected because of the extremely large injection barrier of 1.5–2.5 eV. To obtain efficient electron injection, low work function metals such as calcium (Ca) or aluminum (Al) are selected, however due to their relative ease of oxidation they are not ideal candidates [101, 102]. Another major obstacle is the susceptibility of organic semiconductors to moisture and oxygen under ambient conditions. When a positive gate voltage is applied, organic radical anions may react with moisture or oxygen that has diffused into the film [103]. Therefore many *n*-type transistors can only be operated when processed and measured under encapsulation without moisture and oxygen, leaving them technically unattractive [101].

In order to fabricate reliable *n*-type transistors which can be processed under the same conditions and with the same electrodes as *p*-type transistors, several new organic semiconductors have been designed and synthesized [104, 105]. A number of design rules have been considered for the synthesis of *n*-type semiconducting materials. First, to be able to inject electrons into the LUMO level of semiconductors using environmentally stable electrodes, the LUMO level should be lowered substantially to align with the work function of the stable metal, e.g., gold and increasing the electron affinity of organic semiconductor also improves its environmental stability. Based on the knowledge of *n*-type transistors, it was generally assumed that a high electron affinity (at least 3.0 eV) is required to achieve efficient electron transport in the channel [104]. This can be achieved by taking a known semiconducting core molecule and adding strong electron withdrawn groups such as fluorine, cyano or diimide, and the molecular packing of those molecules may be altered which could improve electron transport in the organic semiconductors. A class of molecules with exceptionally high electron affinity are the fullerenes and their derivatives (e.g. C_{60} and [6,6]-phenyl C_{61}-butyric acid methyl ester (PCBM)). They have been widely demonstrated to obtain electron transport in the transistors with apparently high electron mobilities [106, 107].

Although a high electron affinity improves the environmental stability of the organic semiconductors, the radical anion remains thermodynamically unstable, especially toward oxygen in the presence of moisture [103]. Therefore a kinetic barrier against diffusion of moisture and oxygen into the active channel region during the device operation is required. A very close packing of molecules such as fluorinated copper phthalocyanine (F_{16}CuPc) seems to be beneficial in this respect. Nevertheless, most of the demonstrated *n*-type transistors are unstable under ambient conditions, such as transistors based on fluorinated copper phthalocyanine (F_{16}CuPc), dicyanoperylene-3,4:9,10-bis(dicarboximide) (PDI-8CN_2), and $\alpha - \omega$-diperfluoro-hexyl sexithiophene (DHF-6T) [95, 96, 108]. Only a few examples are demonstrated as air stable *n*-type transistors [109, 110]. In addition to stability and energetic considerations, the orientation of molecules to each other and the associated transfer integral of electrons play a role. Facchetti *et al.* showed that the molecular packing in crystalline semiconductors can strongly affect the observation of *p*- and *n*-type characteristics, and this could also be important for amorphous polymer semiconductors [111, 112]. One of the novel conjugated polymers showing excellent electron transport was developed recently by Yan and Facchetti *et al.* and demonstrated very high electron mobilities up to 0.8 $cm^2\ V^{-1}s^{-1}$ in the *n*-type transistors [99].

3.5.2 *Ambipolar Polymer Field-Effect Transistors*

An ambipolar transistor is one where both electrons and holes can be injected depending on the applied gate voltages. It is like a combination of one *n*-type and one *p*-type transistor. As mentioned above, the complementary technology for fabricating integrated circuits is better compared to the unipolar approach, showing low power dissipation, wide noise margin and high robustness [113]. Unlike in silicon-based technology where *p*- and *n*-type transistors are fabricated by implantation of dopants on nanometer scale, laterally controlled deposition of two different organic semiconductors on the same substrate requires a number of extra process steps, and can significantly increase the manufacturing cost. Consequently, ambipolar materials which can provide both *p*- and *n*-type performance and enable complementary inverters without requiring advanced patterning technique have received high interest and demand [109, 114]. Another advantage of ambipolar-transistor-based inverters is that they work for both positive and negative input voltage and output voltage, depending on the applied voltage, unlike unipolar inverters working only for one charge carrier. However, in such an ambipolar inverter, neither of the transistors is fully switched off and the leakage current could increase power dissipation [101].

The ability to accumulate and transport both electrons and holes in the organic transistor opens new methods to understand of the physics of organic electronic devices. Differences and similarities between electron and hole transport in organic semiconductors and injection mechanism of opposite charge carriers can be investigated by this way. A series of paths have been followed to obtain ambipolar transport in OFETs, and they employ three major directions with respect to the semiconducting layer: bi-layer [115, 116], blend [117, 118], and single component transistors [119–121]. All of these methods have their particular characteristics, advantages, and behave significantly different in their transport physics and current–voltage characteristics. Recently, Zaumseil *et al.* have demonstrated most impressive ambipolar polymer field-effect transistors (PFETs) with a single polymer, e.g., poly(9,9'-dioctylfluorene-co-benzothiadiazole) (F8BT) [101, 120]. The hydrophobic polymer insulators such as polystyrene enable to realize efficient electron transport in the transistor channel of common conjugated polymers without trapping. In particular, their transistors showed excellent light emitting characteristics in the ambipolar region with the high-work-function metal gold thorough top gate/bottom contact configuration. This work will give a deep insight on charge transport and light emitting process in conjugated polymer and heighten possibility to realize electrically pumped conjugated polymer laser. Recently published nice review article will give more details on this story [105].

3.5.3 *All Printed Polymer Field-Effect Transistors*

Inkjet printing is a very attractive technique for the fabrication of functional polymer device. However, in order to make full use of the simplicity of inkjet printing to make PFETs, it is desirable to eliminate the preceding patterning step and improve the pattern resolution. A simple way for the high-resolution patterning solely by inkjet printing is self-aligned printing (SAP) process which has been developed by Sele *et al.* [122]. In this technique the surface of a first functional ink pattern is modified during or after printing, resulting in a hydrophobic surface layer. The subsequently printed ink of a second functional material is repelled by the low surface energy first pattern, and dries in close proximity to the initially deposited ink on the surrounding hydrophobic substrate surface (Fig. 3.10). Thus the SAP decreases the realizable feature size to the nanometer scale whilst simultaneously greatly simplified processing. For instance, no precise alignment is required between the deposition of the first and second pattern. Thanks to the dewetting of ink droplets from a hydrophobic confining area, a submicron gap of conductive inks is

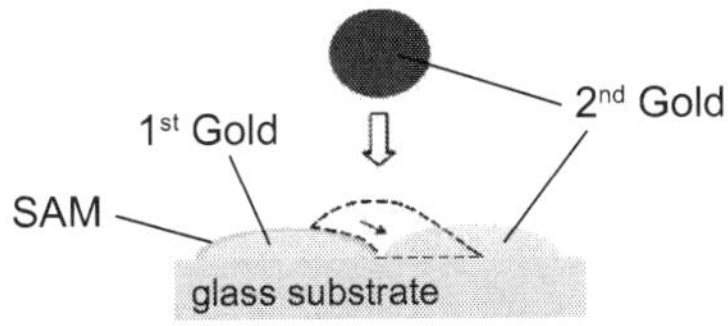

Figure 3.10 Schematic of the self-aligned printing (SAP) technique. A low surface energy layer is defined on a first printed line. Subsequently ink is deposited overlapping the initial line and is forced to recede from the first line due to the hydrophobic surface and created the gap.

formed automatically during the printing process. High-resolution, sub-100 nm channels can be patterned without any lithographic step and the fabricated transistors and invertors using the printed channel can be switched at frequencies up to 40 kHz. However, the low conductivity of conductive polymer, i.e., PEDOT:PSS limits the output currents in the submicron channel. Following that Zhao *et al.* improved the SAP approach by substituting the conducting polymer with dispersed gold nanoparticles (Au NPs), which allowed to obtain higher conductivity of printed electrodes after curing [123]. This improved the performance of the short channel transistors. Finally, fully downscaled printed PFETs have been demonstrated by Noh *et al.* [124]. They reported record high switching speed of 1.6 MHz with high-performance *p*-type conjugated polymer pBTTT. Very small channel length, a low contact resistance, and a low overlap capacitance between gate to source–drain electrode enabled to realize such high-speed circuits with conjugate polymers. This paper stresses the importance of overlap capacitance on the speed of printed polymer circuits.

3.6 Conclusion

In this chapter we presented the theoretical background and some recent progress of PFETs. The performance of PFETs have been highly improved for last decades through development of high-performance conjugated polymer, controlling crystalline of polymer films and organic semiconductor–gate dielectric interface, and optimizing fabrication process. In particular, newly developed high-performance *n*-type conjugate polymers enable to realize high-speed CMOS based integrated circuits providing low energy consumption and a large noise margin. Now state-of-art PFETs show enough performances being able to meet requirements for a few low-end applications. Furthermore, conjugated polymers are considered as a very suitable candidate for semiconductive ink materials due to their high viscosity and better processablitliy making ink formulation for various low-cost graphic art printing techniques such as inkjet, gravure, and screen printing. It is expected that many researches on applying conjugated polymers to roll-to-roll manufacturing processes based on graphic art printing techniques as a semiconductive ink to realize various applications including PFETs will be progressed in near future.

References

1. M. Pope, H. Kallmann, and P. Magnante (1963). Electroluminescence in organic crystals. *J. Chem. Phys.* **38**, 2042–2043.
2. W. Helfrich and W. G. Schneider (1965). Recombination radiation in anthracene crystals. *Phys. Rev. Lett.* **7**, 229–231.
3. C. K. Chiang, C. R. Fincher, Y. W. Park, A. J. Heeger, H. Shirakawa, E. J. Louis, S. C. Gau, and A. G. Macdiarmid (1977) Electrical conductivity in doped polyacetylene. *Phys. Rev. Lett.* **39**, 1098–1101.
4. J. H. Burroughes, D. D. C. Bradley, A. R. Brown, R. N. Marks, K. Mackay, R. H. Friend, P. L. Burns, and A. B. Holmes (1990). Light-emitting diodes based on conjugated polymers. *Nature* **347**, 539–541.
5. R. H. Friend, R. W. Gymer, A. B. Holmes, J. H. Burroughes, R. N. Marks, C. Taliani, D. D. C. Bradley, D. A. Dos Santos, J. L. Bredas, M. Logdlund, and W. R. Salaneck (1999). Electroluminescence in conjugated polymers. *Nature* **397**, 121–128.
6. J. J. M. Halls, C. A. Walsh, N. C. Greenham, E. A. Marseglia, R. H. Friend, S. C. Moratti, and A. B. Holmes (1995). Efficient photodiodes from interpenetrating polymer networks. *Nature* **376**, 498–500.
7. H. Sirringhaus, P. J. Brown, R. H. Friend, M. M. Nielsen, K. Bechgaard, B. M. W. Langeveld-Voss, A. J. H. Spiering, R. A. J. Janssen, E. W. Meijer, P. Herwig, and D. M. de Leeuw (1999). Two-dimensional charge transport in self-organized, high-mobility conjugated polymers. *Nature* **401**, 685–688.

8. G. H. Gelinck, H. E. A. Huitema, E. van Veenendaa, E. Cantatore, L. Schrijnemakers, J. B. P. H. van der Putten, T. C. T. Geuns, M. Beenhakkers, J. B. Giesbers, B.-H. Huisman, E. J. Meijer, E. M. Benito, F. J. Touwslager, A. W. Marsman, B. J. E. van Rens, and D. M. de Leeuw (2004). Flexible active-matrix displays and shift registers based on solution-processed organic transistors. *Nat. Mater.* **3**, 106–110.
9. H. Sirringhaus (2005). Device physics of solution-processed organic field-effect transistors. *Adv. Mater.* **17**, 2411–2425.
10. J. G. Laquindanum, H. E. Katz, A. J. Lovinger, and A. Dodabalapur (1996). Morphological origin of high mobility in pentacene thin-film transistors. *Chem. Mater.* **8**, 2542–2544.
11. D. J. Gundlach, Y. Y. Lin, T. N. Jackson, S. F. Nelson, and D. G. Schlom (1997). Pentacene organic thin-film transistors-molecular ordering and mobility. *IEEE Elec. Dev. Lett.* **18**, 87–89.
12. D. Fichou (2000). Structural order in conjugated oligothiophenes and its implications on opto-electronic devices. *J. Mat. Chem.* **10**, 571–588.
13. Z. Bao, A. Dodabalapur, and A. J. Lovinger (1996) Soluble and processable regioregular poly(3-hexylthiophene) for thin film field-effect transistor applications with high mobility. *Appl. Phys. Lett.* **69**, 4108–4110.
14. B. S. Ong, Y. Wu, P. Liu, S. Gardner (2004). High-performance semiconducting polythiophenes for organic thin-film transistors. *J. Am. Chem. Soc.* **126**, 3378–3379.
15. I. McCulloch, M. Heeney, C. Bailey, K. Genevicius, I. MacDonald, M. Shkunov, D. Sparrowe, S. Tierney, R. Wagner, W. Zhang, M. L. Chabinyc, R. J. Kline, M. D. McGehee, M. F. Toney (2006). Liquid-crystalline semiconducting polymers with high charge-carrier mobility. *Nat. Mater.* **5**, 328–333.
16. H. Sirringhaus, and R. J. Wilson, R. H. Friend, M. Inbasekaran, W. Wu, E. P. Woo, M. Grell, and D. D. C. Bradley (200). Mobility enhancement in conjugated polymer field-effect transistors through chain alignment in a liquid-crystalline phase. *Appl. Phys. Lett.* **77**, 406–408.
17. J. Veres, S. Ogier, G. Lloyd, and D. de Leeuw (2004). Gate insulators in organic field-effect transistors. *Chem. Mater.* **16**, 4543–4555.
18. S. F. Nelson, Y.-Y. Lin, D. J. Gundlach, and T. N. Jackson (1998). Temperature-independent transport in high-mobility pentacene transistors. *Appl. Phys. Lett.* **72**, 1854–1856.
19. G. Horowitz, X. Peng, D. Fichou, F. Garnier (1990). The oligothiophene-based field-effect transistor: how it works and how to improve it. *J. Appl. Phys.* **67**, 528–532.

20. T. Minakata and Y. Natsume (2005). Polyaniline/TiO_2 microspheres prepared by a template-free method. *Synth. Met.* **153**, 1–5.
21. N. Stingelin-Stutzmann, E. Smits, H. Wondergem, C. Tanase, P. Blom, P. Smith, and D. de Leeuw (2005). Organic thin-film electronics from vitreous solution-processed rubrene hypereutectics. *Nat. Mater.* **4**, 601–606.
22. H. Sirringhaus, N. Tessler, and R. H. Friend (1998). Integrated optoelectronic devices based on conjugated polymers. *Science* **280**, 1741–1744.
23. G. Wang, J. Swensen, D. Moses, A. J. Heeger (2003). Increased mobility from regioregular poly(3-hexylthiophene) field-effect transistors. *J. Appl. Phys.* **93**, 6137–6141.
24. R. J. Kline, M. D. McGehee, E. N. Kadnikova, J. Liu, and J. M. J. Fréchet (2003). Controlling the field-effect mobility of regioregular polythiophene by changing the molecular weight. *Adv. Mater.* **15**, 1519–1522.
25. A. Zen, J. Pflaum, S. Hirschmann, W. Zhuang, F. Jaiser, U. Asawapirom, J. P. Rabe, U. Scherf, and D. Neher (2004). Effect of molecular weight and annealing of poly(3-hexylthiophene)s on the performance of organic field-effect transistors. *Adv. Funct. Mater.* **14**, 757–764.
26. R. Peierls (1955). *Quantum Theory of Solids*. Oxford University Press, Oxford.
27. W. P. Su, J. R. Schrieffer, and A. J. Heeger (1979). Solitons in polyacetylene. *Phys. Rev. Lett.* **42**, 1698–1701.
28. P. W. Atkins (1994). *Physical Chemistry*, 5th ed Oxford University Press, Oxford.
29. M. Pope and C. E. Swenberg (1982). *Electronic Processes in Organic Crystals*. Clarendon Press, Oxford.
30. J. Cornil, D. Beljonne, and J. L. Bredas (1995). Nature of optical-transitions in conjugated oligomers. *J. Chem. Phys.* **103**, 834–841.
31. Y. Y. Deng and H. Sirringhaus (2005). Optical absorptions of polyfluorene transistors. *Phys. Rev.* B **72**, 045207.
32. C. D. Dimitrakopoulos and P. R. L. Malenfant (2002). Organic thin film transistors for large area electronics. *Adv. Mater.* **14**, 99–117.
33. L.-L. Chua, J. Zaumseil, J.-F. Chang, E. C.-W. Ou, P. K.-H. Ho, H. Sirringhaus, and R. H. Friend (2005). General observation of *n*-type field-effect behaviour in organic semiconductors. *Nature* **434**, pp. 194–199.

34. G. Horowitz, R. Hajlaoui, H. Bouchriha, R. Bourguiga, M. Hajlaoui (1998). The concept of threshold voltage in organic field-effect transistors. *Adv. Mater.* **10**, 923–927.

35. A. R. Brown, C. P. Jarrett, D. M. de Leeuw, and M. Matters (1997). Field-effect transistors made from solution-processed organic semiconductors. *Synth. Met.* **88**, 37–55.

36. G. Horowitz, R. Hajlaoui, and P. Delannoy (1995). *J. Phys. III France* **5**, 355.

37. J.-F. Chang H. Sirringhaus M. Giles, M. Heeney, and I. McCulloch (2007). Relative importance of polaron activation and disorder on charge transport in high-mobility conjugate polymer field-effect transistors. *Phys. Rev.* B **76**, 205204.

38. A. Salleo, T. W. Chen, A. R. Völkel, Y. Wu, P. Liu, B. S. Ong, and R. A. Street (2004). Intrinsic hole mobility and trapping in a regioregular poly(thiophene). *Phys. Rev.* B **70**, 115311.

39. L. Bürgi, T. J. Richards, R. H. Friend, and H. Sirringhaus (2003). Close look at charge carrier injection in polymer field-effect transistors. *J. Appl. Phys.* **94**, 6129–6137.

40. A. B. Chwang and C. D. Frisbie (2000). Field effect transport measurements on single grains of sexithiophene: role of the contacts. *J. Phys. Chem.* B **104**, 12202–12209.

41. N. Stutzmann, R. H. Friend, and H. Sirringhaus (2003). Self-aligned, vertical-channel, polymer field-effect transistors. *Science* **299**, 1881–1884.

42. J. Zaumseil, K. W. Baldwin, and J. A. Rogers (2003). Contact resistance in organic transistors that use source and drain electrodes formed by soft contact lamination. *J. Appl. Phys.* **93**, 6117–6124.

43. E. J. Meijer, G. H. Gelinck, E. van Veenendaal, B.-H. Huisman, D. M. de Leeuw, and T. M. Klapwijk (2003). Scaling behavior and parasitic series resistance in disordered organic field-effect transistors. *Appl. Phys. Lett.* **82**, 4576–4578.

44. C. Y. Chen and J. Kanicki (1997). Gated-four-probe a-Si:H TFT structure: A new technique to measure the intrinsic performance of a-Si:H TFT. IEEE Elec. *Dev. Lett.* **18**, 340–342.

45. P. V. Pesavento, R. J. Chesterfield, C. R. Newman, and C. D. Frisbie (2004). Gated four-probe measurements on pentacene thin-film transistors: Contact resistance as a function of gate voltage and temperature. *J. Appl. Phys.* **96**, 7312–7324.

46. T. J. Richards and H. Sirringhaus (2007). Analysis of the contact resistance in staggered, top-gate organic field-effect transistors. *J. Appl. Phys.* **102**, 094510.
47. H. Ishii, K. Sugiyama, E. Ito, and K. Seki (1999). Energy level alignment and interfacial electronic structures at organic metal and organicorganic interfaces. *Adv. Mater.* **11**, 605–625.
48. M. A. Abkowitz and H. A. Mizes (1995). Emission limited injection by thermally assisted tunneling into a trap-free transport polymer. *Appl. Phys. Lett.* **66**, 1288–1290.
49. V. I. Arkhipov, E. V. Emelianova, Y. H. Tak, and H. Bässler (1998). Charge injection into light-emitting diodes: Theory and experiment. *J. Appl. Phys.* **84**, 848–856.
50. J. C. Scott and G. G. Malliaras (1999). Charge injection and recombination at the metal-organic interface. *Chem. Phys. Lett.* **299**, 115–119.
51. A. Kahn, N. Koch, and W. Y. Gao (2003). Electronic structure and electrical properties of interfaces between metals and pi-conjugated molecular films. *J. Poly. Sci. B.* **41**, 2529–2548.
52. B. de Boer, A. Hadipour, M. M. Mandoc, T. van Woudenbergh, and P. W. M. Blom (2005). Tuning of metal work functions with self-assembled monolayers. *Adv. Mater.* **17**, 621–625.
53. B. H. Hamadani, D. A. Corley, J. W. Ciszek, J. M. Tour, and D. Natelson (2006). Controlling charge injection in organic field-effect transistors using self-assembled monolayers. *Nano. Lett.* **6**, 1303–1306.
54. N. Zhao, (2008). Field-Effect Transistors Based on Microcrystalline Conjugated Polymers. PhD Thesis, University of Cambridge.
55. H. Klauk (2006). *Organic Electronics: Materials, Manufacturing and Applications*. Wiley-VCH, Weinheim.
56. S. J. Zilker, C. Detcheverry, E. Cantatore, and D. M. de Leeuw (2001). Bias stress in organic thin-film transistors and logic gates. *Appl. Phys. Lett.* **79**, 1124–1126.
57. A. R. Brown (1997) Field-effect transistors made from solution-processed organic semiconductors. *Synth. Met.* **88**, 37–55.
58. H. Klauk, M. Halik, U. Zschieschang, G. Schmid, W. Radlik, and W. Weber (2002). High-mobility polymer gate dielectric pentacene thin film transistors. *J. Appl. Phys.* **92**, 5259–5263.
59. M. J. Powell, C. van Berkel, A. R. Franklin, S. C. Deane, and W. I. Milne (1992). Defect pool in amorphous-silicon thin-film transistors. *Phys. Rev.* B **45**, 4160–4170.

60. A. Salleo and R. A. Street (2003). Light-induced bias stress reversal in polyfluorene thin-film transistors. *J. Appl. Phys.* **94**, 471–479.

61. R. A. Street, A. Salleo, and M. L. Chabinyc (2003). Bipolaron mechanism for bias-stress effects in polymer transistors. *Phys. Rev.* B **68**, 085316.

62. T. N. Ng, J. A. Marohn, and M. L. Chabinyc (2006). Comparing the kinetics of bias stress in organic field-effect transistors with different dielectric interfaces. *J. Appl. Phys.* **100**, 084505.

63. T. Richards and H. Sirringhaus (2008). Bias-stress induced contact and channel degradation in staggered and coplanar organic field-effect transistors. *Appl. Phys. Lett.* **92**, 023512–023514.

64. D. Knipp, R. A. Street, A. Völkel, and J. Ho (2003). Pentacene thin film transistors on inorganic dielectrics: Morphology, structural properties, and electronic transport. *J. Appl. Phys.* **93**, 347–355.

65. P. Stallinga, H. L. Gomes, F. Biscarini, M. Murgia, and D. M. de Leeuw (2004). Electronic transport in field-effect transistors of sexithiophene. *J. Appl. Phys.* **96**, 5277–5283.

66. H. L. Gomes, P. Stallinga, F. Dinelli, M. Murgia, F. Biscarini, and D. M. de Leeuw (2004). Bias-induced threshold voltages shifts in thin-film organic transistors. *Appl. Phys. Lett.* **84**, 3184–3186.

67. J. E. Northrup and M. L. Chabinyc (2003). Gap states in organic semiconductors: Hydrogen- and oxygen-induced states in pentacene. *Phys. Rev.* B **68**, 041202.

68. A. Miller and E. Abrahams (1960). Impurity conduction at low concentrations. *Phys. Rev.* **120**, 745–755.

69. N. F. Mott (1969). Localized states in a pseudogap and near extremities of conduction and valence bands. *Philos. Mag.* **19**, 835–852.

70. S. Roth (1991). *Hopping Transport in Solids*. Elsevier Science, Netherland.

71. D. Monroe (1985). Hopping in exponential band tails. *Phys. Rev. Lett.* **54**, 146–149.

72. M. C. J. M. Vissenberg and M. Matters (1998). Theory of the field-effect mobility in amorphous organic transistors. *Phys. Rev. B* **57**, 12964–12967.

73. H. Bässler (1993). Charge transport in disordered organic photoconductors a monte carlo simulation study. *Phys. Status Solidi B* **175**, 15–56.

74. A. Dieckmann, H. Bässler, and P. M. Borsenberger, (1993). An assessment of the role of dipoles on the density-of-states function of disordered molecular solids. *J. Chem. Phys.* **99**, 8136–8141.

75. L. B. Schein, A. Peled, and D. Glatz (1989). The electric field dependence of the mobility in molecularly doped polymers. *J. Appl. Phys.* **66**, 686–692.
76. P. M. Borsenberger and J. Shi (1995). Hole transport in a vapor deposited phenylenediamine molecular glass. *Phys. Status Solidi B* **191**, 461–469.
77. M. A. Abkowitz (1992). Electronic transport in polymers. *Philos. Mag. B* **65**, 817–829.
78. S. V. Novikov and A. V. Vannikov (1995). Cluster Structure in the Distribution of the Electrostatic Potential in a Lattice of Randomly Oriented Dipoles. *J. Phys. Chem.* **99**, 14573–14576.
79. S. V. Novikov, D. H. Dunlap, V. M. Kenkre, P. E. Parris, and A. V. Vannikov (1998). Essential role of correlations in governing charge transport in disordered organic materials. *Phys. Rev. Lett.* **81**, 4472–4475.
80. L. B. Schein (1992). Comparison of charge transport models in molecularly doped polymers. *Philos. Mag. B* **65**, 795–810.
81. C. Tanase, E. J. Meijer, P. W. M. Blom, and D. M. de Leeuw (2003). Org. *Electron.* **4**, 33–37.
82. L. B. Schein, C. B. Duke, and A. R. McGhie (1978). Observation of the band-hopping transition for electrons in naphthalene. *Phys. Rev. Lett.* **40**, 197–200.
83. J. Cornil, J. P. Calbert, and J. L. Brédas (2001). Electronic Structure of the Pentacene Single Crystal: Relation to Transport Properties. *J. Am. Chem. Soc.* **123**, 1250–1251.
84. L. D. Landau, Phys. Z. Sowjetunion **3**, 664 (1933).
85. H. Fröhlich, H. Pelzer, and S. Zienau (1950). Properties of slow electrons in polar materials. *Philos. Mag.* **41**, 221–242.
86. T. Holstein (1959). Studies of polaron motion: Part I. The molecular-crystal model. *Ann. Phys.* **8**, 325–342.
87. T. Holstein (1959). Studies of polaron motion: Part II. The small polaron. *Ann. Phys.* **8**, 343–389.
88. D. Emin T. Holstein (1969). Studies of small-polaron motion IV. Adiabatic theory of the Hall effect. *Ann. Phys.* **53**, 439–520.
89. P. G. Le Comber and W. E. Spear (1970). Electronic Transport in Amorphous Silicon Films. *Phys. Rev. Lett.* **25**, 509–511.
90. G. Horowitz (1998). Organic field-effect transistors. *Adv. Mater.* **10**, 365–377.

91. G. Horowitz, M. E. Hajlaoui, and R. Hajlaoui (2000). Temperature and gate voltage dependence of hole mobility in polycrystalline oligothiophene thin film transistors. *J. Appl. Phys.* **87**, 4456–4463.

92. A. R. Brown, A. Pomp, C. M. Hart, and D. M. Deleeuw (1995). Logic gates made from polymer transistors and their use in ring oscillators. *Science* **270**, 972–974.

93. G. H. Gelinck, T. C. T. Geuns, and D. M. de Leeuw (2000). High-performance all-polymer integrated circuits. *Appl. Phys. Lett.* **77**, 1487–1489.

94. H. Klauk, M. Halik, U. Zschieschang, F. Eder, G. Schmid, and C. Dehm (2003). Pentacene organic transistors and ring oscillators on glass and on flexible polymeric substrates. *Appl. Phys. Lett.* **82**, 4175–4177.

95. A. Facchetti, M. Mushrush, H. E. Katz, and T. J. Marks (2003). N-type building blocks for organic electronics: A homologous family of fluorocarbon-substituted thiophene oligomers with high carrier mobility. *Adv. Mater.* **15**, 33–38.

96. B. Yoo, B. A. Jones, D. Basu, D. Fine, T. Jung, S. Mohapatra, A. Facchetti, K. Dimmler, M. R. Wasielewski, T. J. Marks, and A. Dodabalapur (2007). High-performance solution-deposited *n*-channel organic transistors and their complementary circuits. *Adv. Mater.* **19**, 4028–4032.

97. H. Yan, Y. Zheng, R. Blache, C. Newman, S. F. Lu, J. Woerle, and A. Facchetti (2008). Solution processed top-gate *n*-channel transistors and complementary circuits on plastics operating in ambient conditions. *Adv. Mater.* **20**, 3393–3398.

98. M. H. Yoon, H. Yan, A. Facchetti, and T. J. Marks (2005). Low-voltage organic field-effect transistors and inverters enabled by ultrathin cross-linked polymers as gate dielectrics. *J. Am. Chem. Soc.* **127**, 10388–10395.

99. H. Yan, Z. H. Chen, Y. Zheng, C. Newman, J. R. Quinn, F. Dotz, M. Kastler, and A. Facchetti (2009). A high-mobility electron-transporting polymer for printed transistors. *Nature* **457**, 679–686.

100. S. Rentenberger, A. Vollmer, E. Zojer, R. Schennach, and N. Koch (2006). UV/ozone treated Au for air-stable, low hole injection barrier electrodes in organic electronics. *J. Appl. Phys.* **100**, 053701.

101. J. Zaumseil, R. H. Friend, and H. Sirringhaus (2006). Spatial control of the recombination zone in an ambipolar light-emitting organic transistor. *Nat. Mater.* **5**, 69–74.

102. B. Q. Sun and H. Sirringhaus (2006). Surface tension and fluid flow driven self-assembly of ordered ZnO nanorod films for high-performance field effect transistors. *J. Am. Chem. Soc.* **128**, 16231–16237.

103. D. M. deLeeuw, M. M. J. Simenon, A. R. Brown, and R. E. F. Einerhand (1997). Stability of *n*-type doped conducting polymers and consequences for polymeric microelectronic devices. *Synth. Met.* **87**, 53–59.

104. C. R. Newman, C. D. Frisbie, D. A. da Silva, J. L. Bredas, P. C. Ewbank, and K. R. Mann (2004). Introduction to organic thin film transistors and design of *n*-channel organic semiconductors. *Chem. Mater.* **16**, 4436–4451.

105. J. Zaumseil and H. Sirringhaus (2007). Electron and ambipolar transport in organic field-effect transistors. *Chem. Rev.* **107**, 1296–1323.

106. T. D. Anthopoulos, B. Singh, N. Marjanovic, N. S. Sariciftci, A. M. Ramil, H. Sitter, M. Colle, and D. M. de Leeuw (2006). High performance n-channel organic field-effect transistors and ring oscillators based on C-60 fullerene films. *Appl. Phys. Lett.* **89**, 213504.

107. P. H. Wobkenberg, D. D. C. Bradley, D. Kronholm, J. C. Hummelen, D. M. de Leeuw, M. Colle, and T. D. Anthopoulos (2008). High mobility *n*-channel organic field-effect transistors based on soluble C-60 and C-70 fullerene derivatives. *Synth. Met.* **158**, 468–472.

108. Z. A. Bao, A. J. Lovinger, and J. Brown (1998). New air-stable *n*-channel organic thin film transistors. *J. Am. Chem. Soc.* **120**, 207–208.

109. T. D. Anthopoulos, S. Setayesh, E. Smits, M. Colle, E. Cantatore, B. de Boer, P. W. M. Blom, and D. M. de Leeuw (2006). Air-stable complementary-like circuits based on organic ambipolar transistors. *Adv. Mater.* **18**, 1900–1904.

110. T. D. Anthopoulos, F. B. Kooistra, H. J. Wondergem, D. Kronholm, J. C. Hummelen, and D. M. de Leeuw (2006). Air-stable *n*-channel organic transistors based on a soluble C-84 fullerene derivative. *Adv. Mater.* **18**, 1679–1684.

111. A. Facchetti, M. H. Yoon, C. L. Stern, H. E. Katz, and T. J. Marks (2003). Building blocks for *n*-type organic electronics: Regiochemically modulated inversion of majority carrier sign in perfluoroarene-modified polythiophene semiconductors. *Angew. Chem., Int. Ed.* **42**, 3900–3903.

112. M. H. Yoon, A. Facchetti, C. E. Stern, and T. J. Marks (2006). Fluorocarbon-modified organic semiconductors: Molecular

architecture, electronic, and crystal structure tuning of arene-versus fluoroarene-thiophene oligomer thin-film properties. *J. Am. Chem. Soc.* **128**, 5792–5801.

113. B. Crone, A. Dodabalapur, Y. Y. Lin, R. W. Filas, Z. Bao, A. LaDuca, R. Sarpeshkar, H. E. Katz, and W. Li (2000). Large-scale complementary integrated circuits based on organic transistors. *Nature* **403**, 521–523.

114. T. Singh, P. Senkarabacak, N. S. Sariciftci, A. Tanda, C. Lackner, R. Hagelauer, and H. Gilles (2006). Organic inverter circuits employing ambipolar pentacene field-effect transistors. *Appl. Phys. Lett.* **89**, 033512.

115. R. B. Ye, M. Baba, Y. Oishi, K. Mori, K. Suzuki (2005). Air-stable ambipolar organic thin-film transistors based on an organic homostructure. *Appl. Phys. Lett.* **86**, 253505.

116. J. Wang, H. B. Wang, X. J. Yan, H. C. Huang, and D. H. Yan (2005). Organic heterojunction and its application for double channel field-effect transistors. *Appl. Phys. Lett.* **87**, 093507.

117. C. Rost, S. Karg, W. Riess, M. A. Loi, M. Murgia, and M. Muccini (2004). Ambipolar light-emitting organic field-effect transistor. *Appl. Phys. Lett.* **85**, 1613–1615.

118. E. J. Meijer, D. M. De Leeuw, S. Setayesh, E. Van Veenendaal, B. H. Huisman, P. W. M. Blom, J. C. Hummelen, U. Scherf, and T. M. Klapwijk (2003). Solution-processed ambipolar organic field-effect transistors and inverters. *Nat. Mater.* **2**, 678–682.

119. T. D. Anthopoulos, C. Tanase, S. Setayesh, E. J. Meijer, J. C. Hummelen, P. W. M. Blom, and D. M. de Leeuw (2004). Ambipolar organic field-effect transistors based on a solution-processed methanofullerene. *Adv. Mater.* **16**, 2174–2179.

120. J. Zaumseil, C. L. Donley, J. S. Kim, R. H. Friend, and H. Sirringhaus (2006). Efficient top-gate, ambipolar, light-emitting field-effect transistors based on a green-light-emitting polyfluorene. *Adv. Mater.* **18**, 2708–2712.

121. E. C. P. Smits, S. Setayesh, T. D. Anthopoulos, M. Buechel, W. Nijssen, R. Coehoorn, P. W. M. Blom, B. de Boer, and D. M. de Leeuw (2007). Near-infrared light-emitting ambipolar organic field-effect transistors. *Adv. Mater.* **19**, 734–738.

122. C. W. Sele, T. von Werne, R. H. Friend, and H. Sirringhaus (2005). Lithography-free, self-aligned inkjet printing with sub-hundred-nanometer resolution. *Adv. Mater.* **17**, 997–1001.

123. N. Zhao, M. Chiesa, H. Sirringhausa, Y. N. Li, and Y. L. Wu (2007). Self-aligned inkjet printing of highly conducting gold electrodes with submicron resolution. *J. Appl. Phys.* **101**, 064513.

124. Y. Y. Noh, N. Zhao, M. Caironi, and H. Sirringhaus (2007). Downscaling of self-aligned, all-printed polymer thin-film transistors. *Nat. Nanotechnol.* **2**, 784–789.

Chapter 4

Polymer Vertical Transistors: Solid-State Vacuum Tubes

Hsiao-Wen Zan
Institute of Physics EO Engineering, National Chiao Tung University, Hsinchu, Taiwan
hsiaowen@mail.nctu.edu.tw

Since the late 19th century, scientists have researched technologies that provide an electronic switch and signal amplifier. The vacuum tube is a triode with a hot cathode, plate anode, and a controlling grid inside an evacuated glass tube as shown in Fig. 4.1. Since its first patent in 1906, the vacuum tube was a dominant technology until the rise of the transistor, around 1960. Yet even today, vacuum tubes survive in specialized areas, including high-power applications, audio amplifiers, and in military applications due to their resistance to electromagnetic pulse. Besides these practical applications, the vacuum tube is elegant in both structure and operation principle, and whether the vacuum tube concept can revive in this era of solid-state electronics, and once again prove its unique advantages, is an interesting consideration.

Polymer Electronics
Edited by Hsin-Fei Meng

ISBN 978-981-4267-84-7 (Hardcover), 978-981-4364-04-1 (eBook)
www.panstanford.com

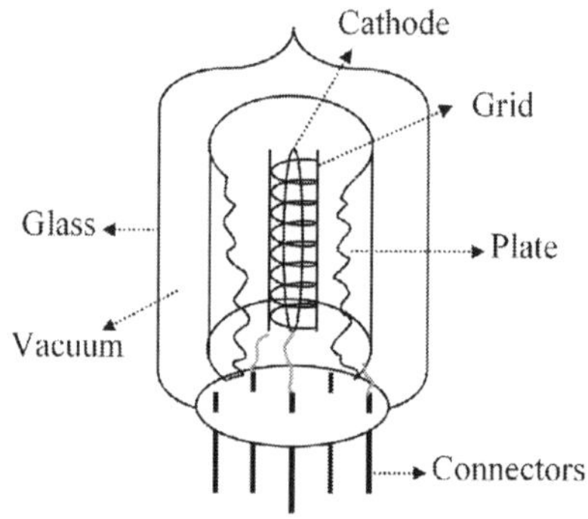

Figure 4.1 The schematic structure of a vacuum tube triode.

Attempts during the 1950s and 1960s to build vacuum tubes using inorganic semiconductors proved unsuccessful because of difficulties with the dopant diffusion process in forming the grid structure [1–5]. Several reports since the 1990s have described research into intrinsic organic semiconductor designs that deliver vacuum tube functionality [6, 7]. From 2007 to 2010, studies successfully demonstrated organic solid-state vacuum tubes with high output current [8, 9], high on/off current ratio [10], and low operation voltage [11]. Their performances are now good enough for real applications. The long journey of the elegant vacuum tube concept therefore opens a promising new chapter a century after its original invention, with the potential of wide technology impact.

This chapter addresses a general comparison between a vertical organic transistor and an organic field-effect transistor in the first section, and introduces various types of vertical organic transistors including the solid-state vacuum tube Then, Section 4.2 describes the fundamental operation principle of an ideal organic solid-state vacuum tube. Sections 4.3 and 4.4 present the realized organic solid-state vacuum tube using a polymer semiconducting material to form a polymer vertical transistor, named the space-charge-limited transistor (SCLT). The transistor includes the fabrication process, carrier injection behavior, leakage control, current improvement, and the grid-field shielding phenomenon.

4.1 Why Use Organic Vertical Transistors?

This section introduces the advantages and challenges of developing organic vertical transistors. When using a vertical structure to achieve low operation voltage easily, leakage control is the key to obtain a high on/off current ratio. This section briefly introduces various types of vertical transistors with different structures, and addresses the revolution of the organic vertical transistor based on the concept of a solid-state vacuum tube.

4.1.1 *Operation Voltage and the On/Off Current Ratio*

As introduced in Chapter 3, conventional organic/polymer field-effect transistors (OFETs) have operation principles similar to metal-oxide-semiconductor field-effect transistors (MOSFETs). For a *p*-type OFET, the transistor channel forms when applying a negative gate bias to bend up the highest occupied molecular orbital (HOMO) in the organic semiconductor close to the semiconductor/insulator interface. Intrinsically, an OFET has a superior gate control and a high on/off current ratio. However, accumulating holes through the gate capacitance achieves the device turn on status. The threshold voltage as well as the operation voltage for an OFET is typically high.

Figure 4.2 denotes the threshold voltages of OFETs reported in leading journals in the past 5 years. Square symbols represent OFETs using small molecules as channel materials [12–16]. Triangle symbols represent polymer FETs [17–21]. Those utilizing high-k materials to serve as gate insulators to enlarge gate capacitance are marked by circle symbols. The threshold voltage for most OFETs is higher than 10 V. OFETs with high-k gate dielectrics such as TiO_2, and ZrO_2 exhibit threshold voltages ranging from 0.7 to 1.1 V [22–24].

Reducing the operation voltage of transistors is crucial for developing integrated circuit or large-area transistor array because low operation voltage gives rise to low power consumption. For complementary metal-oxide-semiconductor (CMOS) technology in very-large-scale integration (VLSI) circuits, using advanced and expensive lithography processes achieves low operation voltage. For example, it is required to realize that MOSFETs operated by 1 V requires the 0.18-μm process. For large-area applications, reducing

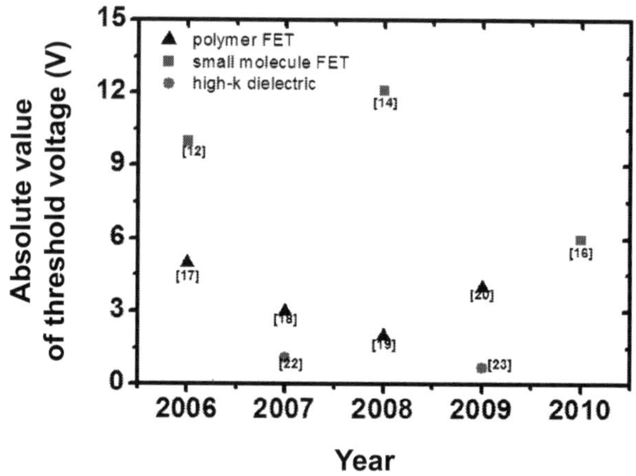

Figure 4.2 The threshold voltages of OFETs in past 5 years.

transistor operation voltage by upgrading lithography system to scale down channel length is not cost effective. Thin-film transistors (TFTs) such as metal-oxide TFTs, amorphous silicon TFTs, and polycrystalline silicon TFTs require operation voltages higher than 5 V. When considering organic or polymer transistors for developing printable, flexible, large-area transistor arrays, operation voltages are typically even higher than 10 V (Table 4.1).

Using the vertical-channel transistor structure with ultra-short channel length is an alternative approach to reduce operation voltage [25–34]. Figure 4.3 plots the operation voltage of various kinds of organic vertical transistors reported in the past 5 years For comparison, data in Fig. 4.2 representing the threshold voltage of OFETs are also denoted in Fig. 4.3 by light grey symbols. Most

Table 4.1 Detail Information of OFETs in past 5 years

	Gate dielectric	Semiconductor	Channel length (μm)	Electrode	Reference
1	PVP SiO_2	Pentacene	90	Au	[12]
2	SiO_2	Polytriarylamines (PTAAs)	10	Al	[18]
3	SiO_2	Pentacene	50–350	Au, Cu	[13]
4	PVP-PMMA	Pentacene	50	Ti	[14]
5	SiO_2 PMMA PC PS	Pentacene	50	Au	[16]

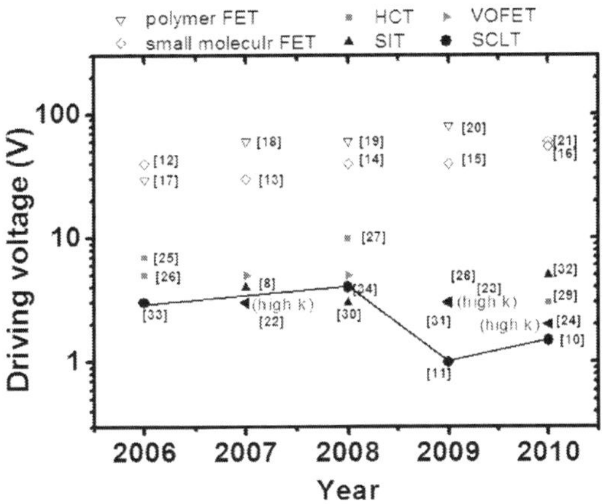

Figure 4.3 The operation voltages of OFETs and various kinds of organic vertical transistors in past 5 years.

organic vertical transistors, with various kinds of structures, have operation voltages lower than 5 V. Some organic vertical transistors can operate at less than 1.5 V [11]. Low operation voltage of the organic vertical transistor is promising for realizing large-area transistor arrays with low power consumption.

The vertical structure, however, typically suffers from a large leakage current, resulting in a poor on/off current ratio. Figure 4.4 plots the on/off current ratio for various kinds of organic vertical transistors reported in the past 5 years. For comparison, Fig. 4.4 also denotes the on/off current ratio of OFETs by light grey symbols. The on/off ratio for most OFETs is higher than 10^5, while that for most organic vertical transistors is lower than 10^3. A key challenge to developing an organic vertical transistor is to raise the on/off ratio to around 10^5, the generally accepted ratio required for practical applications [35]. For example, in the switching transistor for a sensor array with 1000 lines, the current readout is the sum of currents for one line in the on state, and 999 lines in the off state. The latter must be one order of magnitude smaller than the variation range of the former which is typically 10 times. The transistor on-current must therefore be at least 10^5 times larger than the off-current [35]. In 2009, the on/off ratio of the organic vertical transistor rose to 10^4 [11]. In 2010, a polymer vertical transistor

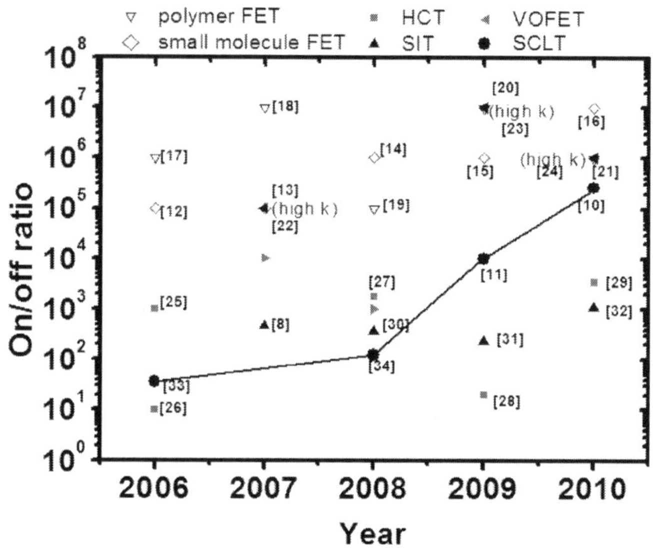

Figure 4.4 The on/off ratios of OFETs and various kinds of organic vertical transistors in past 5 years.

based on the concept of a solid-state vacuum tube, successfully demonstrated an on/off ratio of 3 $\times$ 10^5 [10]. Sections 5.3 and 5.4 introduce obtaining a high on/off ratio in the polymer vertical transistor (Table 4.2).

4.1.2 *Various Organic Vertical Transistors*

This section introduces organic vertical transistors with three different structures. Figure 4.5a shows **Type-A** which utilizes the concept of bipolar junction transistors (BJTs). Previous reports have

Table 4.2 Detailed information of various vertical transistors

Driving voltage (V) Year	HCT	SIT	VOFET	SCLT
2006	5 [26]	4 [45]	-	3 [33]
2007	-	3 [8]	4 [40]	-
2008	10 [27]	2 [30]	5 [39]	4 [34]
2009	3 [28]	3 [31]	-	1.5 [11]
2010	3 [29]	3 [32]	-	1.5 [10]

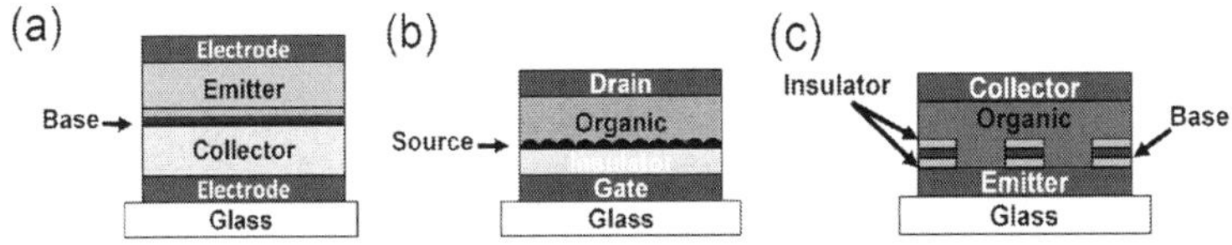

Figure 4.5 Organic vertical transistors with three different structures: (a) Hot-carrier transistor (b) Vertical field-effect transistor (c) Polymer grid triode.

named the device with Structure A hot-carrier transistor (HCT), hot-carrier triode (HCT), or metal-base triode (MBT). Figure 4.5b shows **Type-B** a vertical-type field-effect transistor named the vertical organic field-effect transistor (VOFET). The organic transistor based on the concept of a solid-state vacuum tube has a **Type-C** structure shown in Fig. 4.5c named the polymer-grid triode (PGT) when first reported in 1990 with the grid material as a conducting polymer. Since 1998, it has been called a static-induced transistor (SIT) or a SCLT.

4.1.2.1 Organic vertical transistors based on BJT

A polymer HCT is one of the devices proposed to overcome drawbacks of organic field-effect transistors, such as high operation voltage and low output current. The polymer HCT is a vertical transistor with a carrier perpendicular to the substrate. The thickness of the semiconducting polymer determines the channel length of the device. As in conventional inorganic BJTs, the polymer HCT exhibits two organic diodes connected vertically [36]. An organic diode forms by sandwiching an organic semiconductor layer between two metal layers. When the transistor turns on, the diode between emitter and base (the EB diode) is forwardbiased while the diode between base and collector (the BC diode) is reversedbiased. The carriers inject from the emitter into the base and become hot carriers in the metal because of the large energy barrier at the emitter-base junction The collector will gather most injected carriers as long as the metal base is thinner than the mean free path of the carrier. Figure 4.6 indicates that the ratio of the number of carriers gathered by collector to the one injected from the emitter is

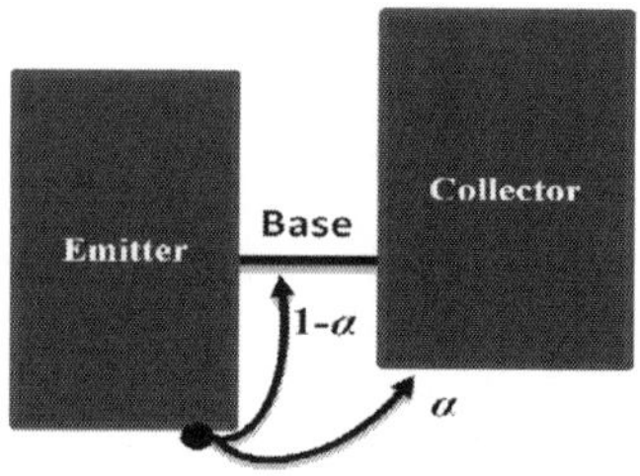

Figure 4.6 The energy band diagram of a hot-carrier transistor.

the transport factor α. Large current amplification results from α are close to one. Current gain β is defined as $\alpha/(1\text{-}\alpha)$.

In 2005, a study reported fabricating a polymer HCT on glass substrate to demonstrate the idea [37]. Figure 4.7a shows the schematic energy band diagram. The study used poly(9-vinylcarbazole) (PVK) and poly(3-hexylthiophene) (P3HT) as the emitter and collector The HOMO for PVK is 6.1 eV and 5.1 eV for P3HT. The energy barrier between base and emitter is 1.8 eV and the one between base and collector is 1 eV. This structure enables the carrier to be far above the Fermi base level while passing through the base and having high kinetic energy to inject into the collector. The study deposited a layer of insulating lithium fluoride (LiF) between emitter and base to further increase the energy barrier at the emitter-base junction. Using the LiF layer weakened mutual dissolution between different polymer layers. Figure 4.7b shows the device characteristics in the common-emitter configuration. The operation voltage was only 5 V, output current density was 0.56 mA/cm^2, and common-emitter current gain was 25.

Although low HOMO of PVK enables a large barrier at the emitter-base junction, it also causes a large carrier injection barrier. Even using gold as the electrode results in an output current density that is not high. Therefore, the study used a P3HT both as emitter and collector in 2008 [27]. Figure 4.8a shows the schematic energy band diagram. The energy offset depends entirely on the LiF layer; therefore, it is essential for this device. Figure 4.8b shows achieving collector current density of 31 mA/cm^2 and β of 31 under an optimized condition, increasing the current density.

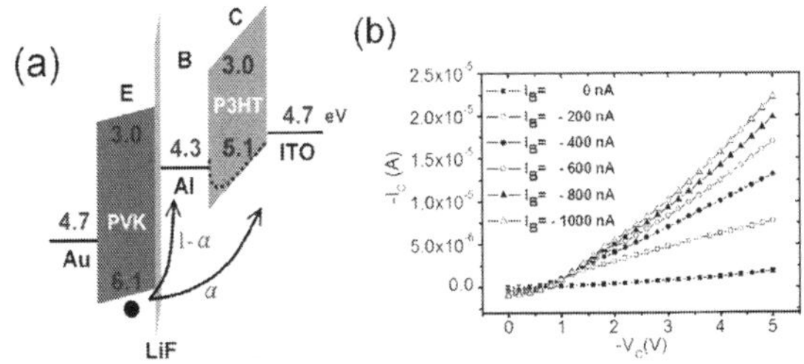

Figure 4.7 (a) The schematic energy band diagram of a hot-carrier transistor with high band-gap emitter (b) Transistor characteristics of a hot-carrier transistor with high band-gap emitter [37].

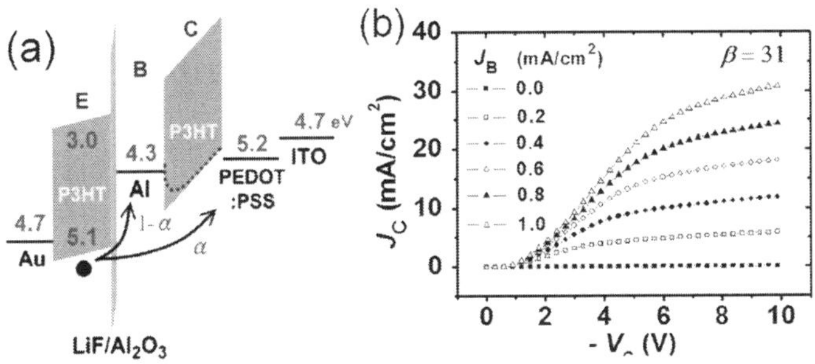

Figure 4.8 The schematic energy band diagram of a hot-carrier transistor with low band-gap emitter (b) Transistor characteristics of a hot-carrier transistor with low band-gap emitter [27].

4.1.2.2 Vertical organic field-effect transistors

Figure 4.9 shows the schematic diagram of the VOFET. Controlling gate bias over channel current relies on the influence of gate bias

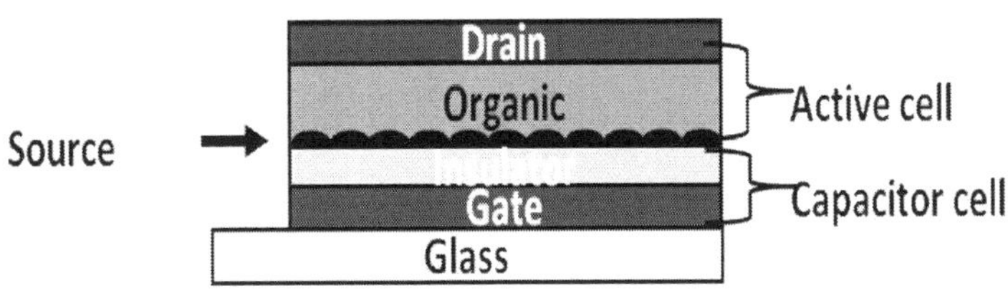

Figure 4.9 The schematic diagram of the vertical organic field-effect transistor.

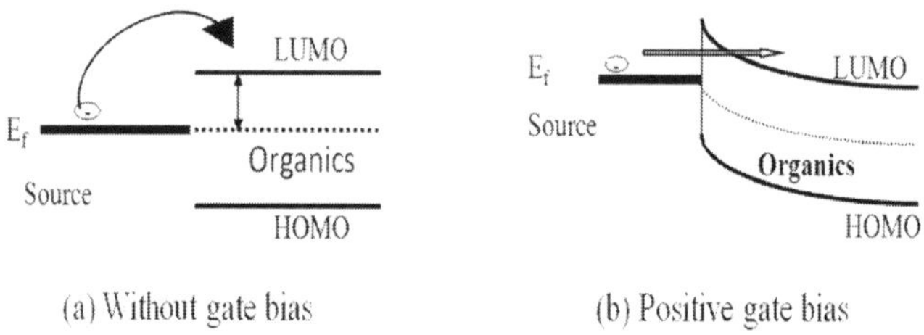

Figure 4.10 The energy band diagram of an *n*-type VOFET (a) when gate bias is zero, and (b) when gate bias is positive.

on the injection barrier between source and organic semiconducting layer. This device, therefore, is normally off Figure 4.10 shows the energy band diagram of an *n*-type VOFET. When gate bias is zero, an injection barrier exists between the source metal and the organic semiconducting layer. When gate bias is positive, it induces negative charges in the source metal through the gate to source capacitance. The negative charges in the source metal elevate the energy band and reduce the effective injection barrier. As a result, electrons inject from the source to the organic layer. With a vertical source-to-drain voltage drop, electrons move from source to drain to form the current. In this device, keeping the leakage current low requires a high injection barrier between the source metal and the organic semiconductor material. The high potential barrier, however, leads to an operation voltage higher than 5 V as in Refs. [38–41]. A low threshold voltage requires a super-high gate capacitance (e.g. 25 $\mu F/cm^2$ in Ref. [38]) to enhance charge accumulation in the source metal.

Table 4.3 lists typical parameters of the reported VOFETs using different channel materials. For VOFET using small-molecule organic material, the on/off current ratio is higher than achieving 10^6. For a polymer VOFET, the on/off current ratio is 10^4, still lower than 10^5.

4.1.2.3 Organic vertical transistors based on solid-state vacuum tubes

Several reports since the 1990s have described research into intrinsic organic semiconductor designs that deliver vacuum tube functionality [6, 7]. The first published in 1994 by Y. Yang and A. J.

Table 4.3 Summary of various polymer grid triodes

	Material in channel	On/off ratio	Working voltage (V)	Current output (A)	Reference
1	C_{60}	10^6	4	10^{-3}	APL, v. 85, p. 5084
2	P3HT	10^4	4	10^{-4}	APL,v.9 3, p. 213301
3	Pentacene	10^4	4	10^{-3}	APL v. 91, p. 083507

Heeger in *Nature* utilizes a porous metallic polymer (polyaniline, PANI) network as the grid material as shown in Fig. 4.11 The self-assembled process forms the grid structure named the polymer grid triode (PGT) [41]. For a *p*-type PGT, holes from the emitter inject (or drain) to the collector (or source) through nanometer holes in the polymer grid. When the metallic polymer potential is negative, holes flow through the polymer grid to arrive at the collector. In this situation, the device operates in the on state. When the metallic polymer potential is positive, the potential inside the holes elevates to form a potential barrier. Holes are not able to cross the barrier; therefore the device is turned off.

The grid made by metallic polymer, however, exhibits low conductivity. As a result, the grid control is weak in PGT. Metal grid fabrication involves depositing metal through a shadow mask with an opening in the micron scale as shown in Fig. 4.12, resulting in poor current density and on/off ratio [42]. Improving grid design using high-density nanometer scale openings increased output current density to greater than 1 mA/cm^2 [8, 34, 43], sufficient for most applications. Realizing high-density nanometer scale openings requires using a spontaneous patterning of higher order nanostructures (SPHOS) process [33, 34, 43, 44–46]. Successful

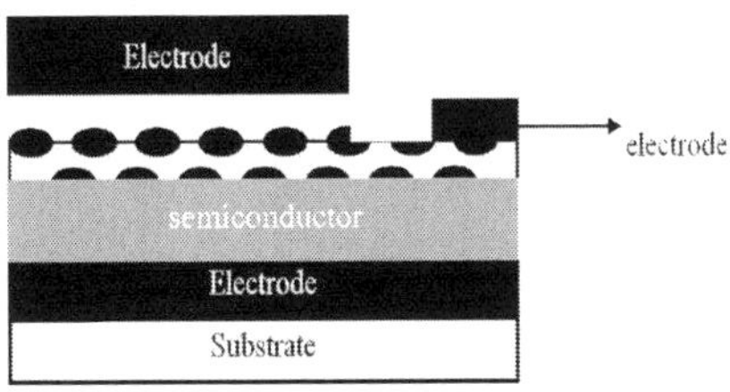

Figure 4.11 The schematic diagram of the polymer grid triode.

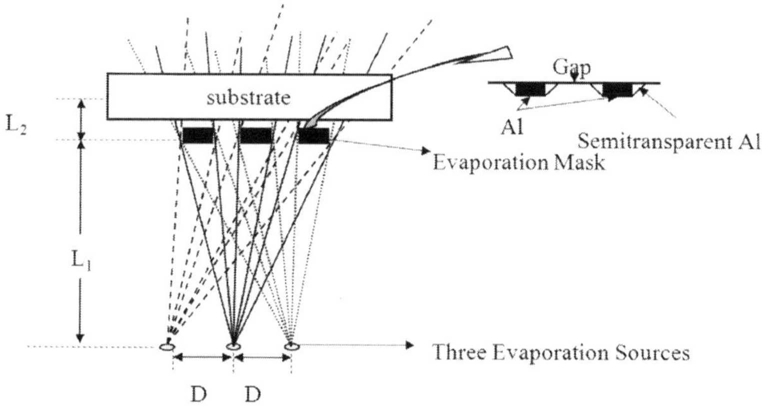

Figure 4.12 Metal grid fabrication involves depositing metal through a shadow mask with an opening in the micron scale.

demonstrations using SPHOS include organic static-induced transistors (OSITs) as shown in Fig. 4.13 and polymer SCLTs as shown in Fig. 4.14. However, for both reports, the on–off current ratio is lower than 1000.

The difficulty regarding the on–off ratio of solid-state vacuum tubes lies in the large spatial overlap between the grid and the collector electrode. In 2009, the on–off ratio of a polymer SCLT rose to 10^4 by introducing an insulator to surround the grid, as shown in Fig. 4.15. In 2010, an on/off ratio of 3×10^5 in polymer SCLT was demonstrated by simply increasing insulator thickness between the

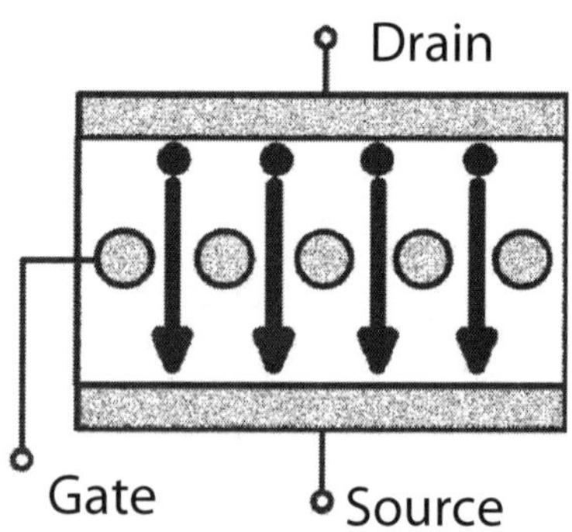

Figure 4.13 The schematic diagram of an organic static-inducedtransistor.

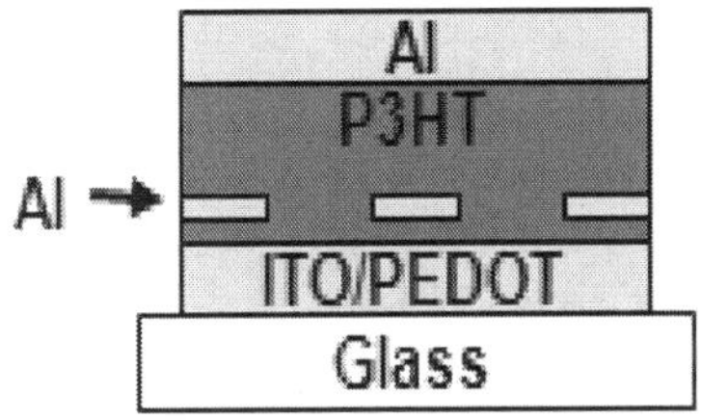

Figure 4.14 The schematic diagram of a polymer space-charge-limited transistor published in 2006.

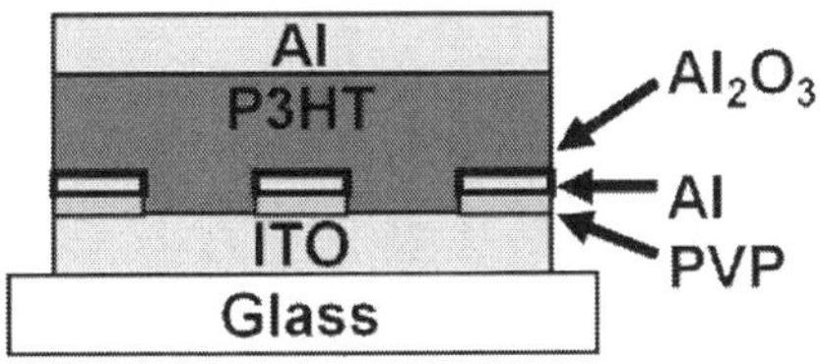

Figure 4.15 The schematic diagram of a polymer space-charge-limited transistor with 60 nm PVP in 2009.

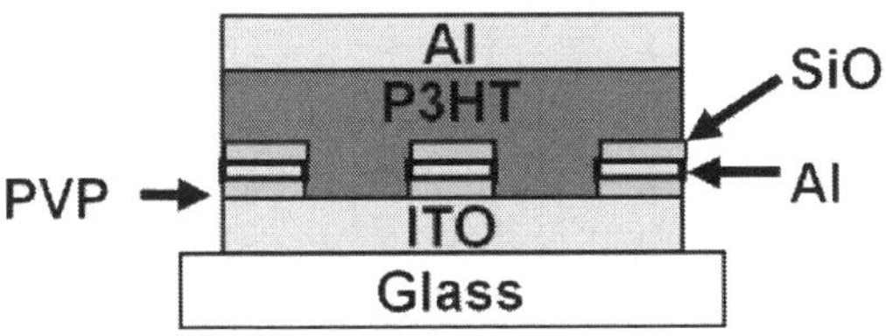

Figure 4.16 The schematic diagram of a polymer space-charge-limited transistor with 60 nm PVP in 2010.

emitter and the grid in Fig. 4.16. Greater thickness allows a wider range of grid potentials to modulate the collector current before breakdown occurs.

With this final improvement, the solid-state vacuum tube offers superior performance features as compared to its field-effect transistor counterpart, with lower voltage, higher output current, and comparable on–off ratio, a performance that is now good enough for real application.

4.2 Operation Principles of Organic Solid-State Vacuum Tubes

This section introduces the operation principles of the polymer vertical transistor based on the concept of the solid-state vacuum tube. Two-dimensional simulations investigate the channel potential profile, the carrier distribution, and the current-voltage characteristics using Silvaco TCAD Atlas simulation software, demonstrating the influences of base bias, carrier mobility and device geometric design on device performance. The space-charge-limited current (SCLC) governs carrier transport in polymer semiconductor material, resulting in the device name of SCLT. The SCLC of the emitter-collector diode is modulated by a metal base electrode embedded within the semiconductor between emitter and collector. The voltages applied on the base and collector determines a potential barrier constituted between emitter and collector. Whether carriers encounter a high or low potential barrier, determines the on and off states of the SCLT. Section 4.2.2.1 discusses the channel potential profile.

4.2.1 *Space-Charge-Limited Current (SCLC) and SCLC Mobility*

Injection of free charge carriers into an intrinsic and trap-free semiconductor from Ohmic metallic contacts results in an inhomogeneous space-charge density. If the current is dominated by drift current and diffusion current is neglected, the current is described as space-charge limited current (SCLC). The SCLC can be derived as below.

The drift current is given by

$$J = e\mu n E \tag{4.1}$$

Start from the equation

$$\frac{\varepsilon_s}{e}\frac{dE}{dx} = n \tag{4.2}$$

and from Eq. (4.1)

$$\frac{\varepsilon_s}{e}\frac{dE}{dx} = \frac{J}{e\mu E} \tag{4.3}$$

$$E\frac{dE}{dx} = \frac{J}{\varepsilon_s \mu} \tag{4.4}$$

$$EdE = \frac{J}{\varepsilon_s \mu} dx \tag{4.5}$$

This differential equation can be integrated to give

$$\frac{1}{2}E^2 = \frac{J}{\varepsilon_s \mu}x + C_1. \tag{4.6}$$

According to the boundary condition E(0) = 0, Eq. (4.6) becomes

$$\frac{1}{2}E^2 = \frac{J}{\varepsilon_s \mu}x \tag{4.7}$$

$$E = \left(\frac{2J}{\varepsilon_s \mu}\right)^{1/2} (x)^{1/2}. \tag{4.8}$$

Therefore

$$V = \int_0^L Edx = \left(\frac{2J}{\varepsilon_s \mu}\right)^{1/2} \frac{L^{3/2}}{3/2} = \left(\frac{8J}{9\varepsilon_s \mu}\right)^{1/2} L^{3/2} \tag{4.9}$$

$$J = \frac{9}{8}\varepsilon_s \mu \frac{V^2}{L^3} = \frac{9}{8}\varepsilon_r \varepsilon_0 \mu \frac{V^2}{L^3} \tag{4.10}$$

Equation (4.10) is referred to as the trap-free square law, the Mott-Gurney square law, and Child's law for solid. This condition is analog to the thermionic vacuum diode, but there are no thermal free carriers and trapping states in the solid. The structure of a polymer SCLT is like an organic diode but with a third electrode inserted inside the polymer to modulate the current. The current in organic diode is commonly described by SCLC, and hence the SCLC is the current that SCLT be modulated.

4.2.2 *Two-Dimensional Simulation*

The simulation uses the device structure shown in Fig. 4.17. The structure is an ideal polymer solid-state vacuum tube. An insulating layer without any leakage current surrounds the base metal grid. An emitter exhibits Ohmic contact with the polymer semiconductor layer. The collector forms Schottky contact with the polymer semiconductor layer. Table 4.4 lists the materials and their

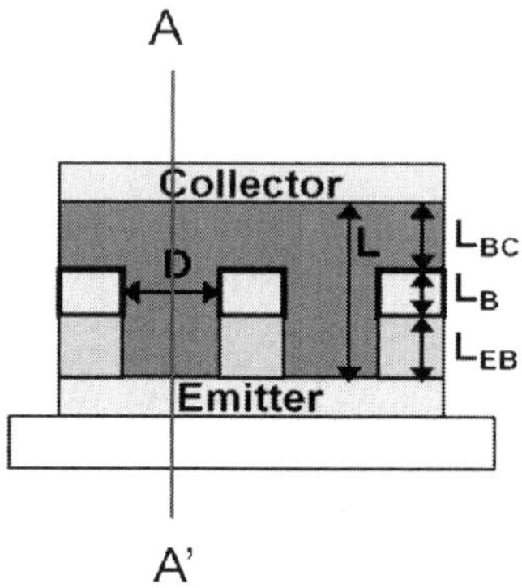

Figure 4.17 The device structure used for simulation. See also Color Insert.

parameters used in the simulation. Indium-tin-oxide (ITO) with oxygen plasma treatment is set as the emitter material. Poly (3-hexylthiophene) (P3HT) is used as a hole transporting polymer semiconductor layer in the simulation. Aluminum is used as the base metal and the collector metal. Parameters used in this section correspond to the parameters in a real polymer SCLT, introduced in Sections 4.3 and 4.4.

Figure 4.17 marks the geometric parameters of this structure and Table 4.4 defines them. D represents the diameter of the holes in the metal grid. L represents the channel length. L_B is the thickness of the metal grid. L_{BE} is the distance between the base metal and the emitter. L_{BC} is the distance between the base metal and collector.

Table 4.4 Simulation parameters

Electrode work function (eV)		Organic polymer	
Emitter	−4.9	Highest occupied molecular orbital (eV)	−5.0
Base	−4.0	Lowest unoccupied molecular orbital (eV)	−3.0
Collector	−4.0	Electron mobility (cm^2/Vs)	10^{-6}
		Hole mobility (cm^2/Vs)	10^{-4}
Geometry		Conductor band density at 300 K (cm^{-3})	5×10^{19}
Channel diameter	100 nm	Valence band density at 300 K (cm^{-3})	5×10^{19}
Channel length	250 nm		
PVP length	100 nm		
Base	40 nm		

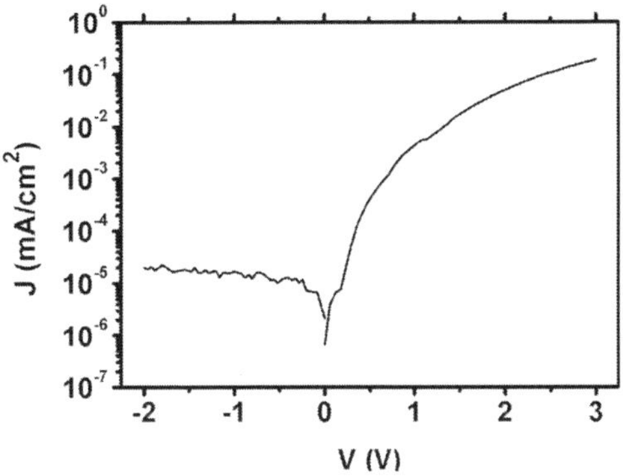

Figure 4.18 The J–V curve of emitter-to-collector diode.

The emitter-to-collector diode (EC diode) exhibits current density-voltage characteristics as shown in Fig. 4.18 when L is 350 nm. Unlike real cases, perfect insulators surrounding the metal grid make the base current equal to zero.

4.2.2.1 Channel potential profile

Figure 4.19 is the simulated channel potential profile from emitter to collector at the central channel. The collector-to-emitter potential (V_{CE}) is −2 V. The x-axis in Fig. 4.19 represents the A–A′ position marked by the red line in Fig. 4.17. When base-to-emitter potential

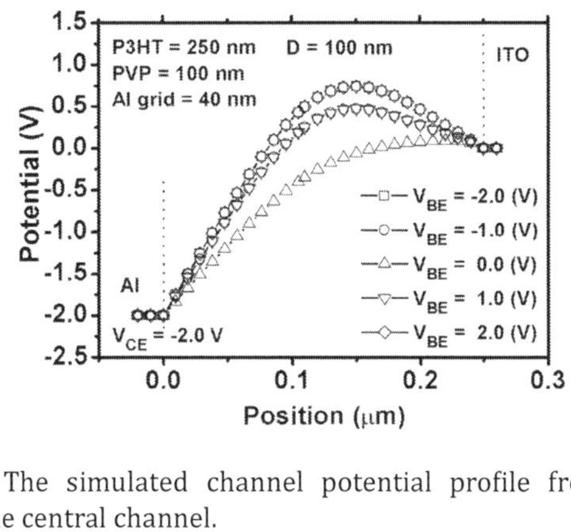

Figure 4.19 The simulated channel potential profile from emitter to collector at the central channel.

(V_{BE}) changes from +2 V to −2 V, the channel potential profile along the A–A′ position is shown in Fig. 4.19. The dashed lines represent a potential profile without potential barrier and the device operates in on state.

The solid lines denote a formed potential barrier to impede hole transport from emitter to collector.

4.2.2.2 Transfer characteristics and the output characteristics

According to the potential profile shown in Fig. 4.19, the corresponding collector current densities (J_C) plotted as a function of V_{BE} (i.e. the transfer characteristics) are depicted in Fig. 4.20 when V_{CE} is -2 V. When V_{BE} is smaller than -0.5 V, the device turns on and J_C is large. When V_{BE} is larger than 0.4 V, the device turns off and J_C is smaller than 0.1 mA/cm^2. The reciprocal of the slope in the transition region from off to on, marked by the dashed line in Fig. 4.20, represents the control ability of the base potential on the channel current. In MOSFET, the reciprocal of the slope in the transition region is named the subthreshold swing ($S.S.$). This chapter borrows the subthreshold swing ($S.S.$) name to represent the base control ability in SCLT. In Fig. 4.20, the ideal on–off current ratio is larger than nine orders.

Figure 4.21 shows the influence of V_{CE} on transfer characteristics. Observations show an almost parallel shift behavior when V_{CE} changes from −1 to −4 V.

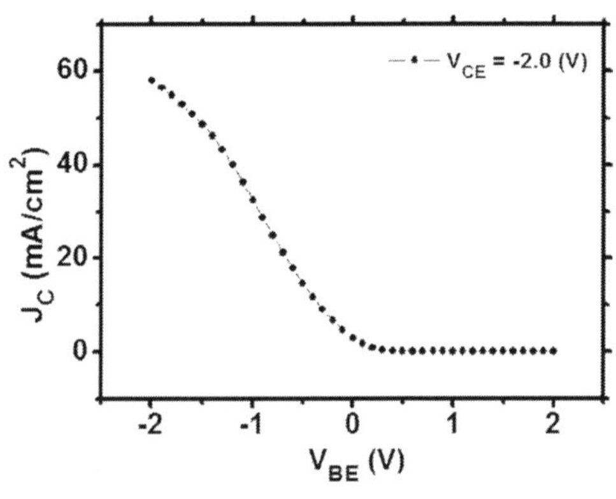

Figure 4.20 The simulated transfer characteristics when V_{CE} is fixed.

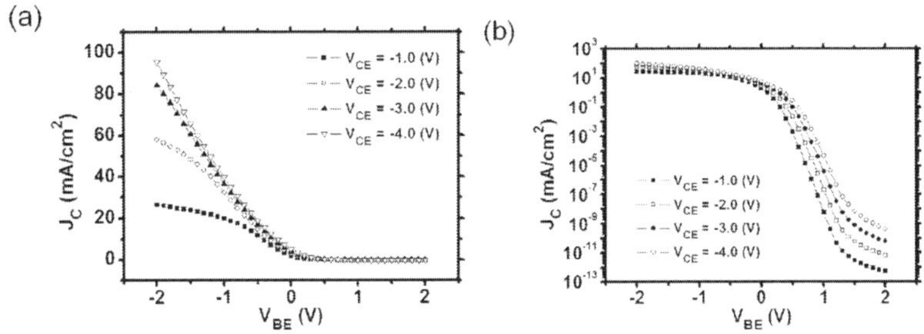

Figure 4.21 The influence of V_{CE} on transfer characteristics.

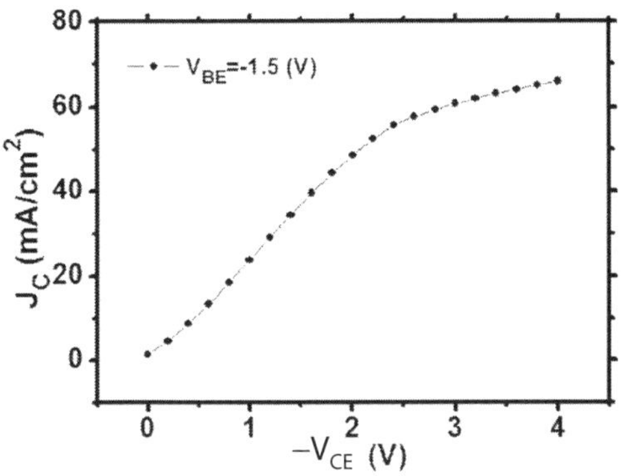

Figure 4.22 The simulated transfer characteristics when V_{BE} is fixed.

The output characteristics, which are the collector current densities (J_C) plotted as a function of V_{CE} are shown in Fig. 4.22 when V_{BE} is -1.5 V. The curve has no saturation behavior. Figure 4.23 shows the output characteristics when the y-axis plots in a log scale and V_{BE} varies from $+2.5$ V to -2.5 V. The SCLT, with current parameters, acting similar to a diode in the on state, causes the lack of saturation behavior. Section 4.2.3.2 obtains saturated J_C–V_{CE} curves by performing a suitable device geometric design.

4.2.3 *Mobility and Geometry Effects*

Based on the SCLC mechanism described in Eq. (4.10), increasing carrier mobility and reducing channel length leads to increased

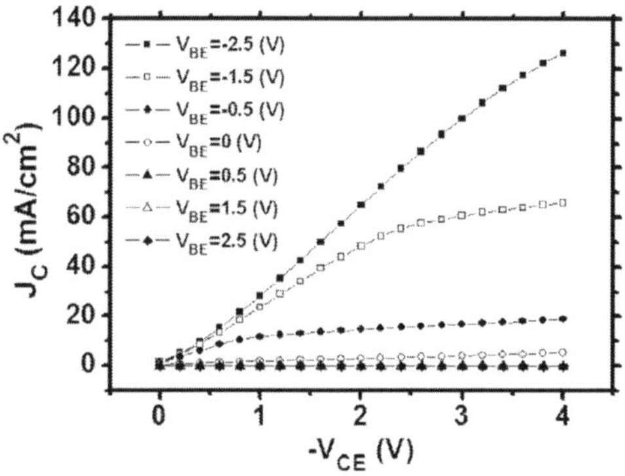

Figure 4.23 The simulated output characteristics at various V_{BE}.

output current. In organic semiconductor materials, SCLC mobility varies from 10^{-6} cm^2/Vs to 10^{-1} cm^2/Vs. For a *p*-type SCLT, injected holes contribute to the channel current while injected electrons form leakage between EC electrodes.

However, in SCLT, the channel potential profile is influenced by V_{CE} as well as V_{BE}. For a classic vacuum tube, the potential at the center of the grid openings is determined by $\lambda V_{BE} + V_{CE}$, where λ is a geometric factor. The influences of hole mobility, electron mobility, and geometric design on SCLT performance are discussed.

Table 4.5 Simulation parameters for understanding the influence of carrier mobility

Electrode work function (eV)		Organic polymer	
Emitter	−4.9	Highest occupied molecular orbital (eV)	−5.0
Base	−4.0	Lowest unoccupied molecular orbital (eV)	−3.0
Collector	−4.0	Conductor band density at 300 K (cm^{-3})	5×10^{19}
		Valence band density at 300 K (cm^{-3})	5×10^{19}
Geometry			
Channel diameter	100 nm		
Channel length	250 nm		
PVP length	100 nm		
Base	40 nm		

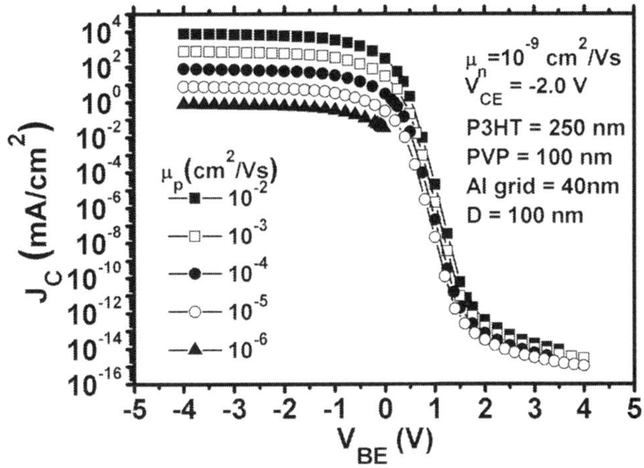

Figure 4.24 The simulated transfer characteristics.

4.2.3.1 Mobility effects

For a given ideal SCLT with parameters given in Table 4.5, increasing hole mobility leads to increased current as shown in Fig. 4.24. The off-state current remains almost unchanged when hole mobility changes from 10^{-6} to 10^{-2} cm^2/Vs. On the contrary, off current varies obviously when electron mobility increases from 10^{-6} to 10^{-3} cm^2/Vs. The off current behavior implies that, in an ideal case, the hole current in the off state is much lower than the electron current. As shown in Figs. 4.25 and 4.26, electrons injected from collector contribute to the ideal off-state current.

For real cases, however, the parasitic leakage current exists. Insulators surrounding the base electrode are not perfect and the base current is not zero. The leakage current is typically much higher than the ideal off current. Observing the ideal off current generated by injected electrons is difficult in real cases.

4.2.3.2 Geometry effects

The current research investigates the influences of geometric parameters including opening diameters (D), channel length (L), metal grid thickness (L_B), and the relative position of base to channel (L_{BE}/L_{BC}) by changing one parameter at a time to observe

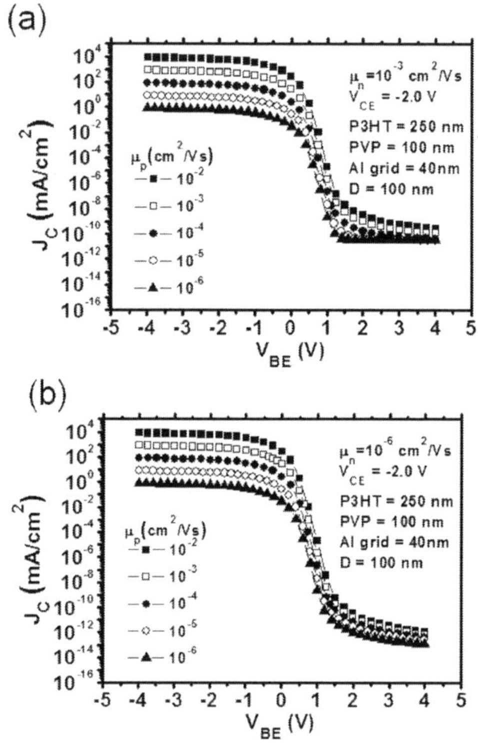

Figure 4.25 (a) The simulated transfer characteristics when electron mobility is 10^{-3} cm^2/Vs (b) The simulated transfer characteristics when electron mobility is 10^{-6} cm^2/Vs.

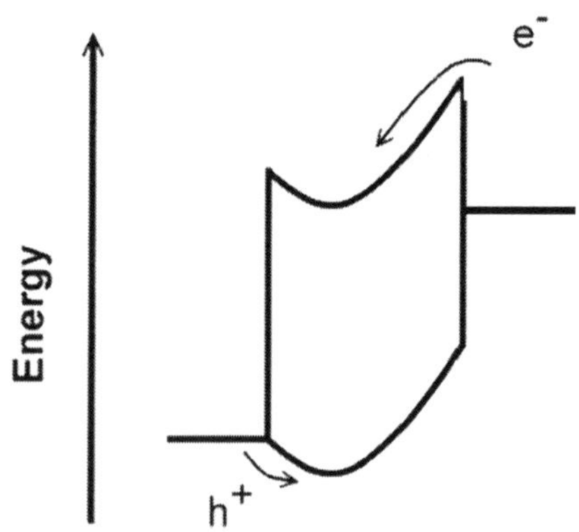

Figure 4.26 Electron may be injected from the collector.

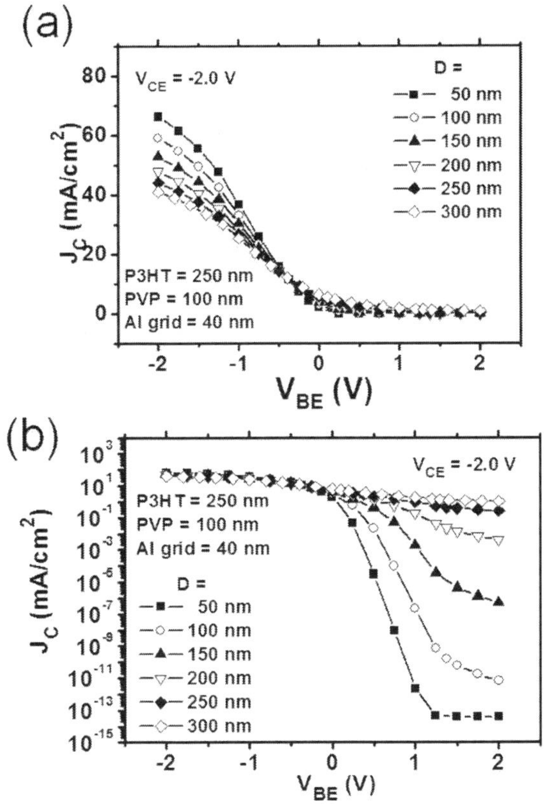

Figure 4.27 The influence of opening diameter *D* on the transfer characteristics.

its influences on transfer characteristics. Figure 4.27 shows that reducing opening diameters (D) significantly improves the base control and reduces $S.S.$ For $D = 50$ nm, $S.S.$ is as low as 76.92 mV/dec. This $S.S.$ value is close to the theoretical limit of the subthreshold swing in the silicon MOSFET, which is 60 mV/dec.

Reducing channel length (L) while keeping the base in the center of the channel ($L_{\rm BE}/L_{\rm BC} = 1\ /\ 1$) leads to reduced current as shown in Fig. 4.28. Increasing metal grid thickness ($L_{\rm B}$) also significantly improves base control, as shown in Fig. 4.29. Moving the relative position of the base in the upper channel ($L_{\rm BE}/L_{\rm BC} = 2/1$ or $4/1$) or lower channel ($L_{\rm BE}/L_{\rm BC} = 1/2$ or $1/4$) leads to a loss of base control.

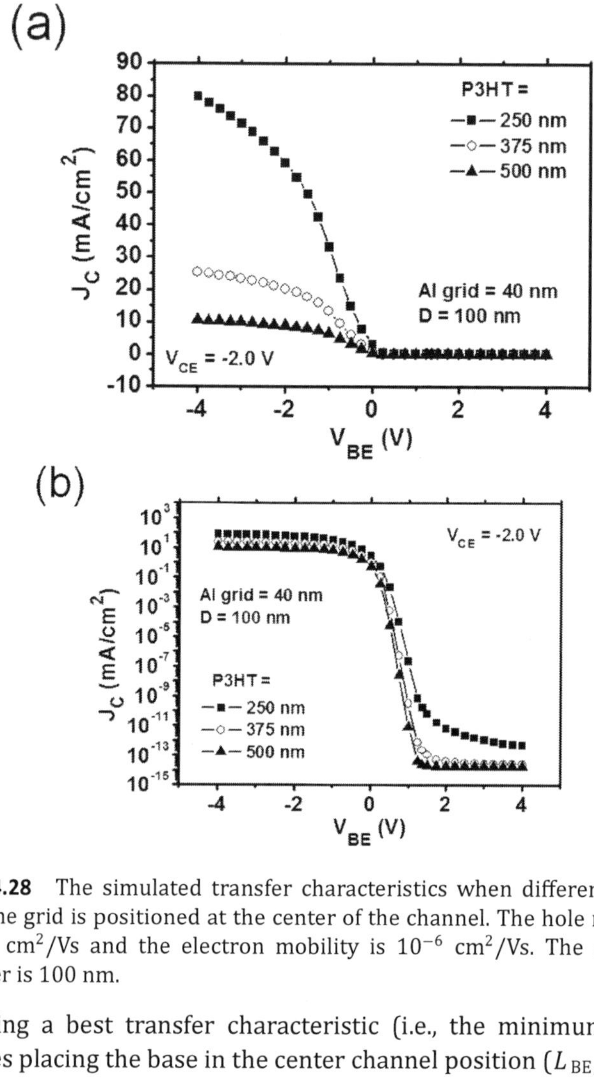

Figure 4.28 The simulated transfer characteristics when different L are used. The grid is positioned at the center of the channel. The hole mobility is 10^{-4} cm^2/Vs and the electron mobility is 10^{-6} cm^2/Vs. The opening diameter is 100 nm.

Obtaining a best transfer characteristic (i.e., the minimum $S.S.$) requires placing the base in the center channel position ($L_{\mathrm{BE}}/L_{\mathrm{BC}} = 1/1$).

In solid-state vacuum tube, the geometric arrangement also affects the saturation behavior of the channel current. Two ultimate situations are observed.

1. When the base grid is placed close to the emitter as shown in Fig. 4.30a, the J_{C}–V_{CE} plot is shown in Fig. 4.30b. Both the x-axis

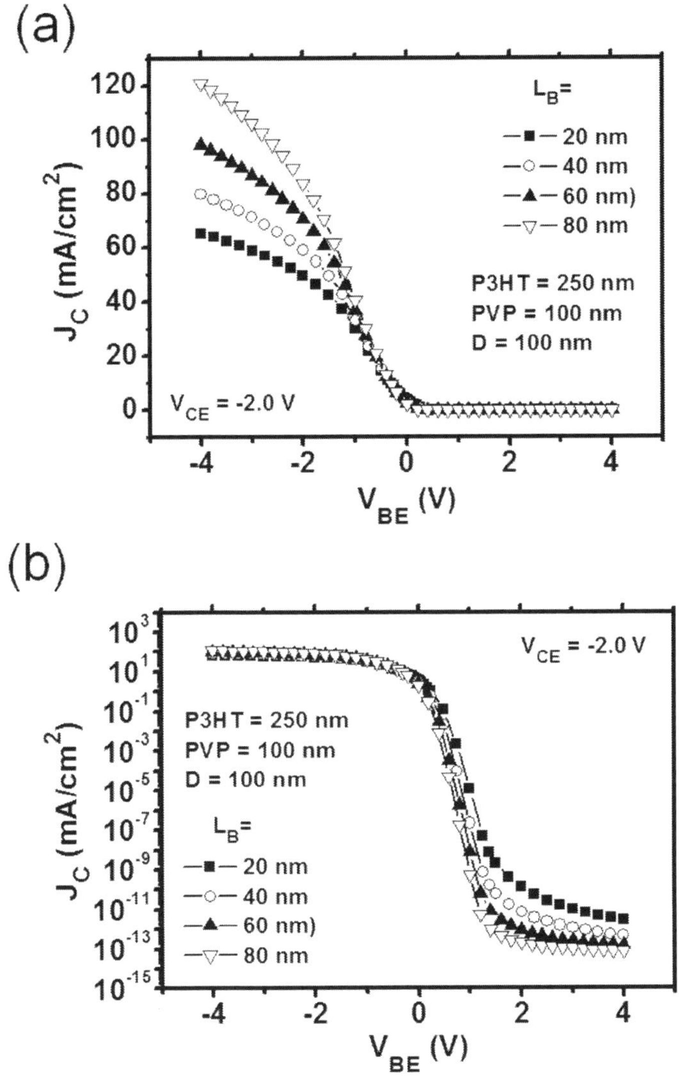

Figure 4.29 The influence of base thickness L_B on the transfer characteristics; L is 250 nm and PVP is 100 nm.

and y-axis are in linear scales in Fig. 4.30b and are in log sales in Fig. 4.30c. A current-voltage behavior dominated by the SCLC relationship is observed, indicating that the current is mostly

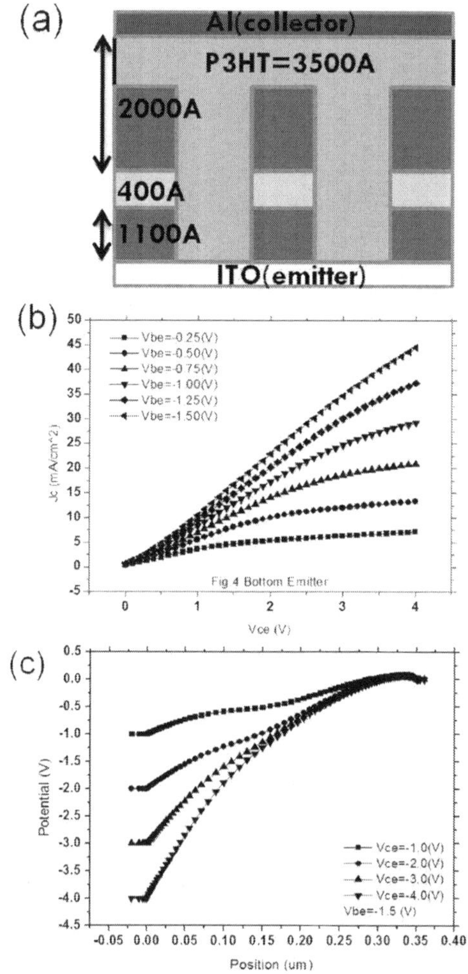

Figure 4.30 (a) The structure of a SCLT with base grid placed close to the collector (b) The output characteristics in linear scale (c) The output characteristics in log scale.

governed by V_{CE} rather than V_{BE}. The J_C–V_{CE} characteristics are not saturated. If the device is applied in circuit, the nonsaturated characteristics cause a very small output resistance, which is usually unwanted. For the inverter application, the nonsaturated characteristics also cause poor voltage swing in the voltage

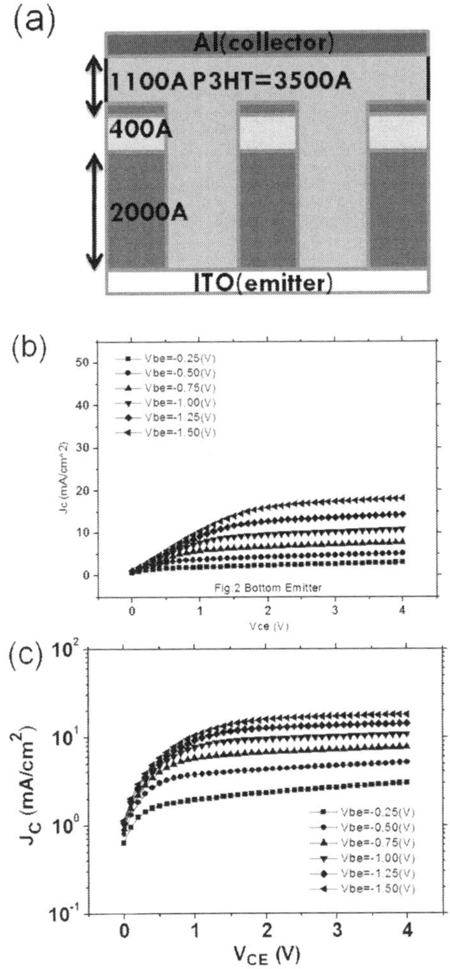

Figure 4.31 (a) The structure of a SCLT with base grid placed close to the emitter (b) The output characteristics in linear scale (c) The potential profiles.

transfer characteristic. For conventional vacuum-tube circuit, double or even triple grids are required to enhance the base control and to obtain the saturation behavior. In SCLT, the saturation behavior can be obtained by putting the base grid close to the collector as shown in Fig. 4.31b.

2. When the base grid is placed close to the collector as shown in Fig. 4.31a, the J_C–V_{CE} plot is shown in Fig. 4.31b. Both the x-axis and y-axis are in linear scales. A well-saturated behavior is obtained. The potential profile as shown in Fig. 4.31c explains the saturation behavior. In SCLC, the channel current is mostly influenced by the potential profile from emitter to the openings surrounded by grid metal. When base grid is placed close to the collector and far from the emitter, the potential at openings is less sensitive to collector voltage. The current is controlled mostly by V_{BE} rather than V_{CE}. When V_{CE} increases, it only increases the electric field from openings to collector. However, the bottleneck of the channel current is controlled by the electric field from emitter to openings. As a result, well-saturated J_C–V_{CE} curves are obtained.

4.3 P3HT-Based Space-Charge Limited Transistors (SCLTs): Fabrication Process and Injection Behavior

This section introduces a real polymer vertical transistor as a solid-state vacuum tube. The solid-state version of the vacuum tube triode is named the "polymer space-charge-limited transistor" (SCLT). This device uses P3HT as a hole transporting polymer semiconductor layer. Studies have realized the grid-like base electrode using spontaneous patterning of a higher order nanostructure (SPHOS) process [33, 34, 43, 44–46]. This section introduces the fabrication process to realize SCLT, and then discusses important factors to reduce the injection barrier, control the leakage current and improve the on current of SCLT. Finally, this study investigates the grid-field shielding phenomenon of SCLT.

4.3.1 *Fabrication Process*

Figure 4.32 shows the structure of the P3HT-based SCLT and Fig. 4.33 shows the detailed process flow chart with process steps. The related scanning electron microscopy (SEM) images are also attached.

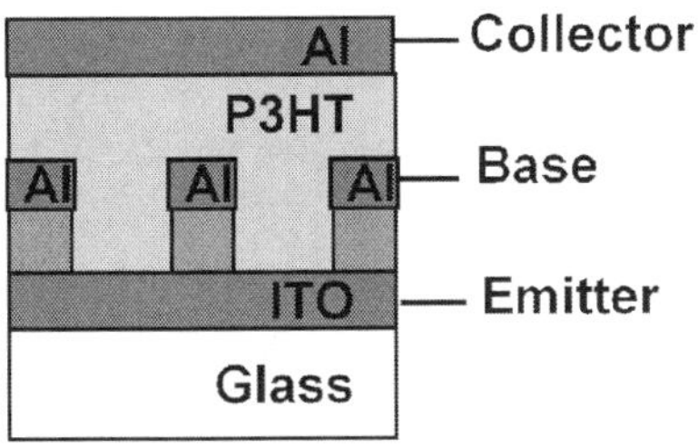

Figure 4.32 The device structure of the P3HT-based SCLT.

Step 1: Preparation of ITO Substrate with Bottom Insulating Layer – The SCLT is fabricated on an ITO glass substrate which is first flattened by 150 W O_2 plasma for 30 min. Cross-linkable poly(4-vinylphenol) (PVP) can be used as the bottom insulator. PVP is spin coated on the ITO substrate and then cross-linked at 200°C for 60 min in vacuum. The thickness of PVP can vary from 60 to 400 nm. As shown in Fig. 4.33, a thicker PVP provides better insulating properties. However, a thicker PVP also increases channel length, which is the thickness of the P3HT layer. In Section 4.3.2, the thickness of PVP is 60 nm. Then, in Sections 4.4.1 and 4.4.2, the thickness of PVP is 210 nm to provide better insulating properties. Poly (melamine-co-formaldehyde) methylated (Aldrich, Mw ~511) is utilized as a cross-linking agent for PVP.

Step 2: Absorption of Self-Organized Spheres – The surface of PVP is turned into hydrophilic by a short time exposure of 50 W O_2 plasma before submerging the substrate into 200 nm positively charged polystyrene (PS) spheres (Merck, K6-020) diluted in ethanol solution with 0.8 wt%. The PS spheres are adsorbed on the PVP surface to serve as the shadow mask. After submerging in a PS sphere solution for 3 min, the substrate is then transferred to a beaker with boiling isopropanol solution for 10 s. The substrate is immediately blown dry to form two-dimensional colloidal arrays.

Step 3: Grid Base Fabrication – Then, 40 nm Al is evaporated as a metal base electrode. In Sections 4.4.1 and 4.4.2, 50 nm SiO is also evaporated sequentially as the upper insulating layer. After removing the PS spheres with an adhesive tape (Scotch, 3M), the

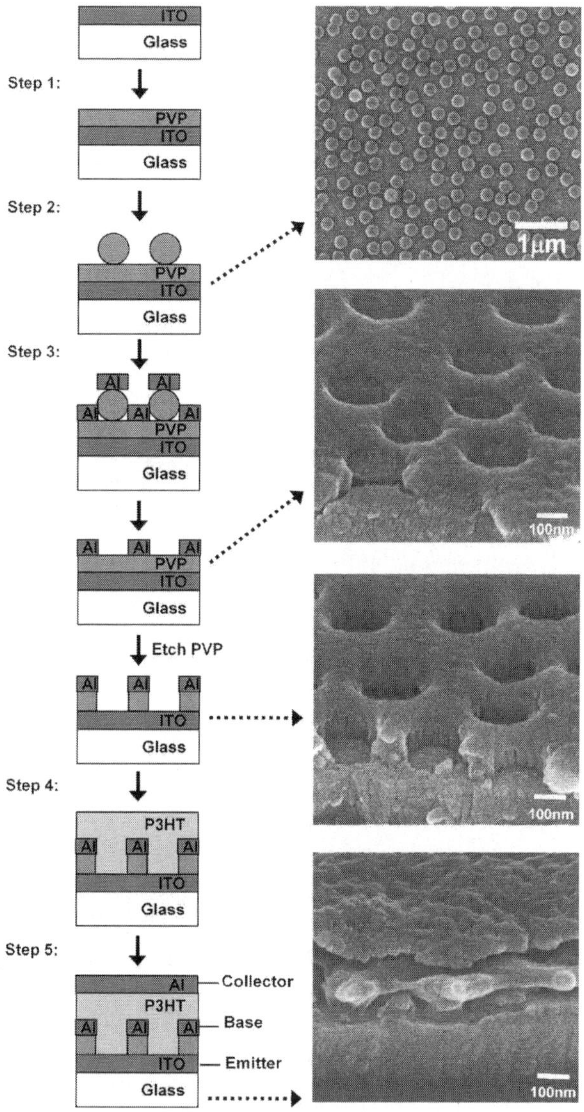

Figure 4.33 Fabrication process flow chart.

PVP at sites without Al coverage is removed by 150 W O_2 plasma treatment. During the O_2 plasma treatment, Al_2O_3 is also grown on the Al base.

Step 4: Polymer Semiconductor Coating – P3HT can be used as a hole transporting polymer semiconductor layer. Different solvents and coating methods can be applied to improve crystallinity as well as hole mobility of P3HT. In Section 4.3.2, P3HT of 1200 Å from 2.5 wt% chlorobenzene is spin coated on samples. In Section 4.4.1, a 350 nm P3HT layer is spin coated onto the substrate from 4.5 wt% chlorobenzene solution. In Section 4.4.2, for the device prepared by spin coating, P3HT in chlorobenzene or dichlorobenzene is spin coated on the substrate until the film dried. As for the device prepared by dichlorobenzene solvent annealing, P3HT in dichlorobenzene (2.5 wt%) is spin coated at 450 rpm for 30 s and then 1200 rpm for 1.2 s to form a P3HT wet film. The wet film is kept still in the glove box until the film dried. The P3HT film is annealed at 140°C for 10 min. In Section 4.5.2, two kinds of active layer were used. One is P3HT of 3500 Å from 4.5 wt% chlorobenzene to form a conventional P3HT SCLT. The other is a blended P3HT/(6,6)-phenyl-C61-butyric acid methyl ester (PCBM) of 3500 Å from 4.5 wt% chlorobenzene to form a new P3HT/PCBM SCLT. The P3HT: PCBM blending ratio is 1:0.1 by weight.

Step 5: Top Metal Evaporation – Finally, the top metal is deposited to complete the SCLT with the active area as 1 mm^2. In Sections 4.3.2, 4.4.1, and 4.4.2, the top metal material is aluminum (Al) to provide a large hole and electron injection barrier between the top metal and P3HT.

4.3.2 *Carrier Injection Behavior*

Since the vertical channel is a thin polymer semiconductor film sandwiched between two metals (emitter and collector), the thin polymer film gives rise to a low channel resistance while the injection barrier from the metal to the polymer causes large contact resistance at the metal/polymer interface. When the operation voltage across the vertical channel reduces, the voltage drop at the metal/polymer interface significantly influences device performance. A reduced voltage drop achieves a vertical polymer transistor operated with ultra-low voltage. At the same time, a low vertical field enables better leakage current control to obtain a high on–off current ratio at low operation voltage.

In Section 4.3.2.1, the Mott–Schottky barrier between metal and organic semiconductor is introduced. This section investigates the injection barrier from indium-tin-oxide (ITO) to P3HT. Introducing oxygen plasma treatment adjusts the ITO work function to improve hole-injection from ITO into P3HT. The improvement, however, suffers from an aging effect. Finally, this study demonstrates P3HT-based SCLTs with improved hole-injection properties The reduced hole-injection barrier at the ITO/polymer interface enhances the hole-current at low voltage. Meanwhile, a low electric field suppresses the leakage current. A collector to emitter voltage (V_{CE}) as low as -0.84 V obtains a high on-off ratio of 10^4.

4.3.2.1 ITO work function adjustment and aging effect

Many reports have demonstrated that changes in surface chemical composition, surface species, and surface carbon contribute to enhanced ITO work function [47–49]. Figure 4.34 shows that enhanced ITO work function results in reduced energy barrier at the interface of ITO and P3HT. Many methods can enhance the ITO work function including UV/ozone and plasma treatment. As shown in Step 3 in Fig. 4.33, the ITO emitter surface is treated with O_2 plasma during the plasma etch-back process before spin coating P3HT. As a result, hole-injection from ITO to P3HT improves with O_2 plasma treatment.

The ITO/P3HT/Al diodes were fabricated on an ITO glass substrate with or without O_2 plasma treatment on ITO. P3HT

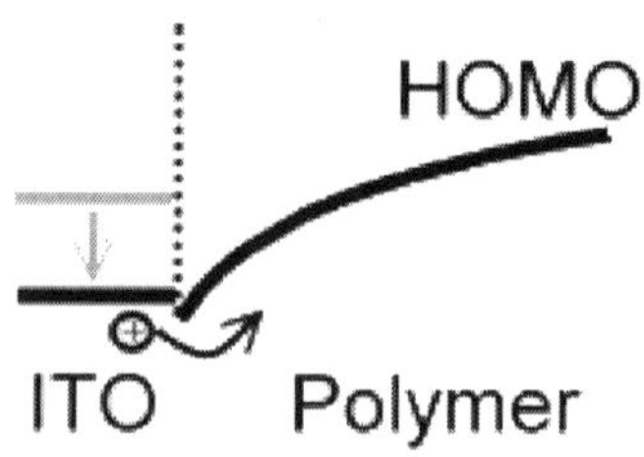

Figure 4.34 Enhanced ITO work function by oxygen plasma.

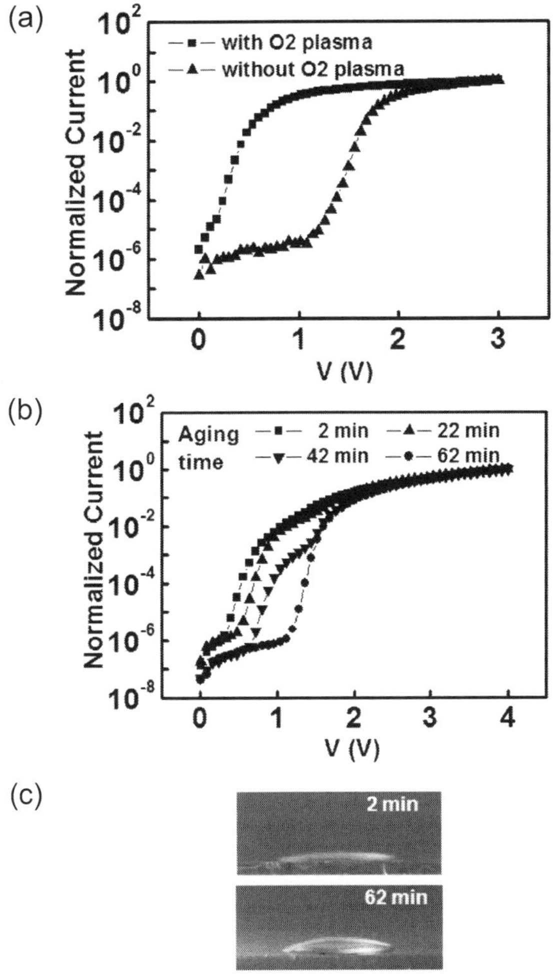

Figure 4.35 (a) Normalized current-voltage curve of P3HT diode (b) Aging effect of the oxygen plasma treated ITO (c) Images of water drop on ITO surface with aging time of 2 and 62 min.

of 1200 Å was spin-coated from 2.5 wt% chlorobenzene. Figure 4.35a shows the normalized current-voltage curves of ITO/P3HT/Al diodes. The normalized current turns on rapidly at low voltage for the diode fabricated on an O_2-plasma-treated ITO. The normalized current turns on at around 1 V for the diode fabricated on ITO without O_2 plasma treatment. For the diode fabricated with O_2-plasma-treated ITO, holes can be injected into P3HT at lower voltage. For the diode with a low hole-injection barrier, hole-injection occurs at low voltage and hence the turn-on voltage is low. As for the diode with a high injection barrier, high voltage is needed for hole-tunneling through the barrier. Since the enhanced ITO work function relates to the change of surface carbon and surface species, aging effect is expected to significantly influence ITO work function.

Figure 4.35b shows investigating aging effect on the O_2-plasma-treated ITO. The ITO/P3HT/Al diodes were fabricated on an ITO glass substrate with or without O_2 plasma treatment on ITO. P3HT of 1200 Å was spin-coated from 2.5 wt% chlorobenzene. To investigate the influence of aging time on diode characteristics, some ITO glass substrates were stored in a glove box for various aging times after O_2 plasma treatment followed by spin coating P3HT. The plasma chamber was connected to the glove box such that O_2-plasma-treated ITO substrates did not contact with ambient air during the aging time investigation. Since 2 min is necessary for transferring samples from the plasma chamber into the glove box and spin-coating P3HT, the minimum aging time was estimated at 2 min. Longer aging time was achieved by storing samples in the glove box for various time periods. For the diode with 2 min aging time, a rapidly increasing current was observed when the forward-bias voltage was applied. However, as the aging time increased from 2 to 62 min, the turn-on voltage of ITO/P3HT/Al diode increased. In the meantime, the contact angle also rose with aging time, shown in the Fig 4.35c. An aging time longer than 62 min caused no apparent change in turn-on voltage and contact angle. The following explains the observation. Increased aging time caused the surface oxygen concentration to decrease and the surface carbon concentration to increase, resulting in decrement of the ITO work function, and increment of the hole-injection barrier, as well as turn-on voltage [50]. After 62 min of aging time, the surface condition returned to before O_2 plasma treatment, and the benefit brought by O_2 plasma treatment disappeared.

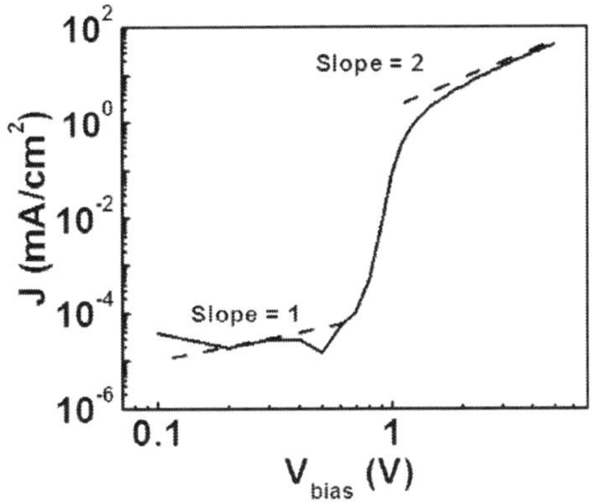

Figure 4.36 Output current density as a function of V_{CE}.

4.3.2.2 P3HT SCLT with ultralow operation voltage

When SCLT is turned on, the relationship between channel current (I_C) and channel voltage (V_{CE}) is observable by the log J_C–log V_{CE} curve, as shown in Fig. 4.36. The region with a slope equal to 2 refers to the current as the SCLC. When the slope is equal to 1, conduction relates to the intrinsic carrier density since the interface barrier blocks hole-injection [51]. Reducing the injection barrier is crucial to obtaining high current under low voltage.

Knowing that O_2 plasma treatment decreases the turn-on voltage and enhances the hole current at low voltage, while aging causes the opposite effect, this study investigates the influence of O_2 plasma treatment and aging effect on SCLT characteristics. The process flow introduced in Section 4.3.1 is the basis for fabricating SCLTs. After Step 3, samples are stored in a glove box for various times (the aging time) before spin coating with P3HT.

Figure 4.37a,b compare the SCLT characteristics made with aging time of 2 min (low-injection-barrier SCLT) and 62 min (high-injection-barrier SCLT). The V_{BE} changes from negative to positive to turn on and turn off the SCLT. The off current reduces by increasing V_{BE} until leakage current occurs to cause a large base current density (J_B). Apparently, V_{BE} can modulate the collector current density (J_C) for both SCLTs. However, the low-injection-barrier SCLT can be turned on at low voltage ($V_{CE} = -0.3$ V), but the high-injection-

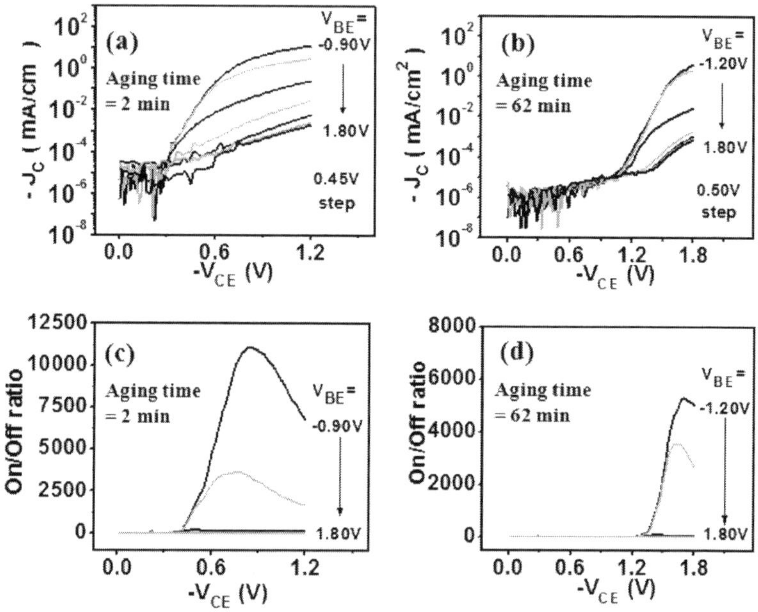

Figure 4.37 The output characteristics of the SCLT with aging time of (a) 2 min and (b) 62 min. The on–off ratio of the SCLT with aging time of (c) 2 min and (d) 62 min.

barrier SCLT can only be turned on with higher voltage (V_{CE} = −1 V). This agrees with the results obtained from Fig. 4.35b. Low-injection-barrier SCLT shows good transistor characteristics and the output current density is around 12.7 mA/cm^2 as $V_{CE} = -1.2$ V. The base current density is in the order of 10^{-4}–10^{-3} mA/cm^2 and hence the current gain $|J_C/J_B|$ is as large as 10^4 to 10^5. Figure 4.37c,d plot the on–off ratio for SCLT with a low and high injection barrier as a function of V_{CE}. A high on–off ratio of 10^4 for low-injection-barrier SCLT is obtained at V_{CE} = -0.84 V, while that for high-injection-barrier SCLT is about 5000 at V_{CE} = -1.69 V. Even when applying more negative V_{CE} and V_{BE} on high-injection-barrier SCLT to turn on the emitter-to-collector diode, the on–off ratio is still lower than that of low-injection-barrier SCLT.

The lower on/off ratio in high-injection-barrier SCLT results from higher base-collector voltage drop, which increases leakage current and reduces the on/off ratio. Operating the SCLT under low voltage can prevent leakage current. A low injection barrier from ITO to P3HT is essential to obtain high output current and high on/off ratio at low operation voltage.

4.4 P3HT-Based SCLTs: Leakage Control and Mobility Improvement

4.4.1 *Leakage Current Control for High On/Off Current Ratio*

Ideally, the SCLT on and off states are determined by whether carriers encounter a high or low potential barrier. Figure 4.38 shows a simulated potential distribution at the central vertical channel when collector-to-emitter voltage (V_{CE}) is –1.5 V and base-to-emitter voltage (V_{BE}) varies from 2.5 to –2.5 V. The figure describes a strong bias effect on potential distribution. A more positive V_{BE} leads to a higher potential barrier, and lower channel current, thus, turning off the channel current. A simulated channel current density J_C vs.

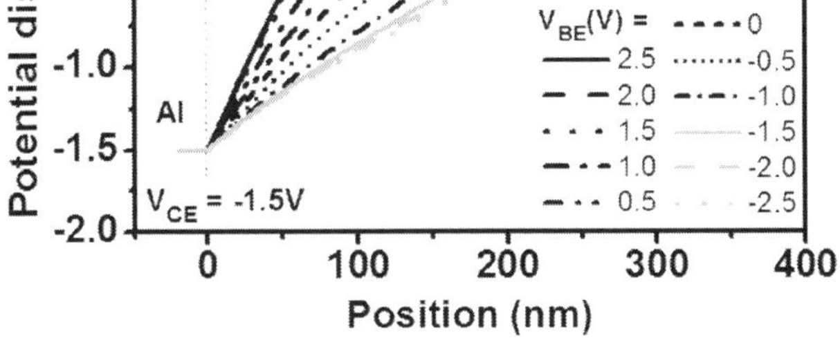

Figure 4.38 The simulated potential distribution at the central vertical channel when collector-to-emitter voltage (V_{CE}) is −1.5 V and base-to-emitter voltage (V_{BE}) varies from 2.5 to −2.5 V.

V_{BE} plot, for V_{CE} equal to -1.5 V reveals that, in the absence of all parasitic leakage, the device on/off current ratio could be greater than nine orders of magnitude.

However, in a real situation, parasitic leakage occurs to elevate the off current significantly. As a result, parasitic leakage control is crucial to achieve high on/off current ratio.

4.4.1.1 Leakage current analysis

Figure 4.39 depicts origins of the SCLT off current. Figure 4.40 shows the corresponding potential profile for the SCLT off-state. At a fixed negative V_{CE}, the output collector current density (J_C) is the sum of J_{C1} and J_{C2}. J_{C1} is the current density from emitter to collector, J_{C2} is the leakage current density from base to collector, and J_3 is the leakage current between base and emitter. The arrows indicate the current direction. Tuning the V_{BE} from negative to positive suppresses J_{C1}; however, J_{C2} and J_3 both increase, since the voltage drop across the base and collector (V_{BC}) and V_{BE} are increased. Reaching the minimum of J_C is possible as J_{C2} becomes comparable to J_{C1}. Hence, to maximize the on/off ratio, while keeping the current gain high, requires methods to reduce J_{C2} and J_3. Previous works used a thin PVP insulating layer 60 nm below the base, and an ultra thin Al_2O_3 insulating layer no thicker than 5 nm above the base [52, 53]; but the properties of these two insulating layers were insufficient for keeping V_{BE} high without increasing J_{C2} and J_3. This

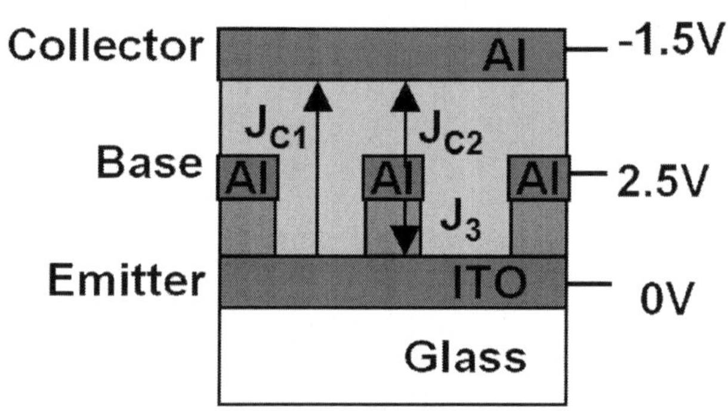

Figure 4.39 The analysis of the origin of the SCLT off current.

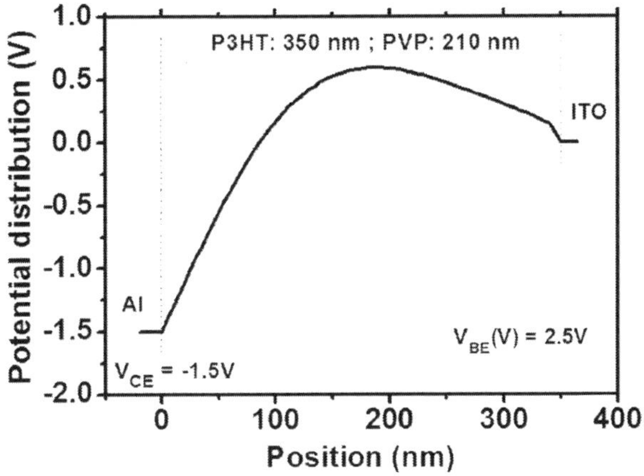

Figure 4.40 The corresponding potential profile for the SCLT off-state.

section demonstrates a polymer vertical transistor with enhanced insulating layer properties, which allows a high V_{BE} and makes a high on/off ratio of 3×10^5 possible. The experiment added a 50 nm insulating SiO layer above the base electrode, and increased the thickness of the PVP insulating layer below the base electrode to 210 nm. P3HT thickness maintained at 350 nm, because reducing P3HT to 270 nm increases the risk of short circuit, while increasing the layer to 430 nm causes a reduction in J_C.

4.4.1.2 Fabrication process and device design

The SCLT was fabricated according to the process flow described in Section 4.3.1. In this section, the thickness of PVP is 210 nm. A 50-nm-thick SiO was evaporated sequentially after base metal (a 40-nm-thick aluminum) evaporation to serve as the top insulator. In a glove box, a 350 nm P3HT layer was spin coated onto the substrate from 4.5 wt% chlorobenzene solution.

Figure 4.41 shows insulating properties of PVP layers of various thicknesses. It records the J–V curves of emitter-base diodes (ITO/PVP/Al) after removing PVP at the site without Al coverage. The measurement process involved applying a voltage to the emitter while keeping the base grounded. The 210 nm PVP device exhibited lower leakage current density than the 60 nm device. For the

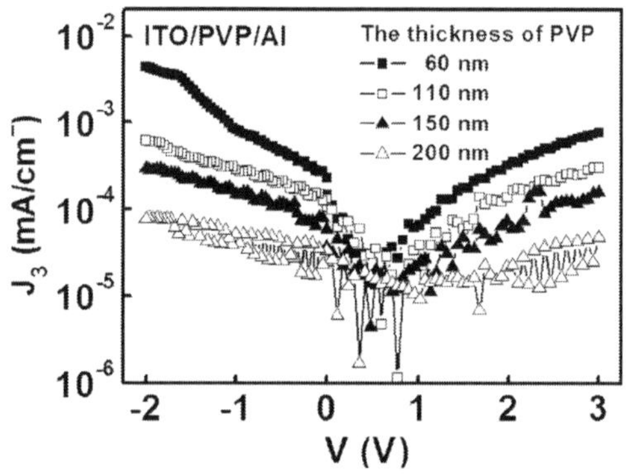

Figure 4.41 The insulating properties of PVP layers of various thicknesses.

210 nm device, J_3 value was in the order of 10^{-5} mA/cm^2 for positive bias below 2 V. The enhanced insulating properties of 210 nm PVP allowed the emitter-base diode to sustain higher bias without breakdown. To improve SCLT characteristics, the P3HT layer should be no thinner than the upper surface of the base and depressions in the P3HT surface should not be observable in SEM images above the base opening sites [54]. Since a 210 nm PVP layer is used, the P3HT layer must be thicker than 210 nm. Thus, diodes in the ITO/P3HT/Al structure with various P3HT thicknesses are fabricated to understand the diode characteristics. The current densities of these diodes were recorded with 2.5 V applied on an ITO emitter, and various carrier mobility were deduced from the currents following SCLC. As shown in Fig. 4.42, current density decreases with increasing P3HT thickness; however, carrier mobility increases with increasing P3HT thickness. This phenomenon was also observed in small molecule organic diodes [55, 56]. The authors concluded that the organic thin film's low mobility originated from an interfacial trap state and molecular arrangement on the surface. In this case, since P3HT is known to be sensitive to surface conditions, the molecular orientation condition in the region close to the ITO must differ from that of the bulk far away from the ITO. The influence of interfacial conditions on the diode characteristics becomes weaker as the thickness of P3HT increases, and hence the mobility of P3HT increases with increasing thickness.

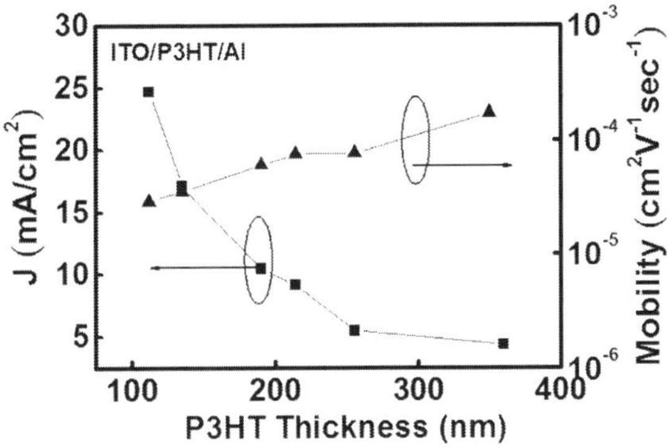

Figure 4.42 The variation of the current density and carrier mobility with increasing P3HT thickness.

4.4.1.3 High on/off ratio characteristics

Figure 4.43 shows SCLT transistor characteristics, with J_C values plotted as a function of V_{CE} and V_{BE}. Figure 4.43c shows the transfer characteristics of SCLT determined from Fig. 4.43a. This work grounded the ITO emitter and negatively biased the Al collector at V_{CE}. If V_{CE} is equal to -1.5 V, then when V_{BE} is -1.1 V, the SCLT is in the on state, and J_C is 1.71 mA/cm^2. When V_{BE} is 2.5 V, J_C suppresses and the SCLT is in the off state. Base current density (J_B) is in the order of 10^{-4} to 10^{-5} mA/cm^2, and hence the current gain $|J_C/J_B|$ is as large as 10^4 to 10^5. Figure 4.43b shows the on/off ratio and current gain of SCLT as functions of V_{CE}. Dividing J_C (V_{BE} at -1.1 V or -0.5 V) by J_C (V_{BE} at 2.5 V) obtains the on/off ratio, and dividing J_C by J_B when V_{BE} is -1.1 V or -0.5 V obtains the current gain. The V_{CE} at -1.5 V achieves the maximum on/off ratio of 3×10^5, and the current gain is still greater than 1000. This on/off ratio is larger than the one performed in Section 4.3, and is the highest value so far reported for polymer vertical transistors. Since the transistor on current is in the same order as in Section 4.3, the improved on/off ratio must arise from the reduced off current, which may come from two sources. First, applying a more positive V_{BE} voltage to the base can improve the on/off ratio. As compared with the V_{BE} value of 1.8 V used in Section 4.3, the V_{BE} of 2.5 V used here effectively suppresses the off current. Higher V_{BE} values do help to improve the on/off ratio

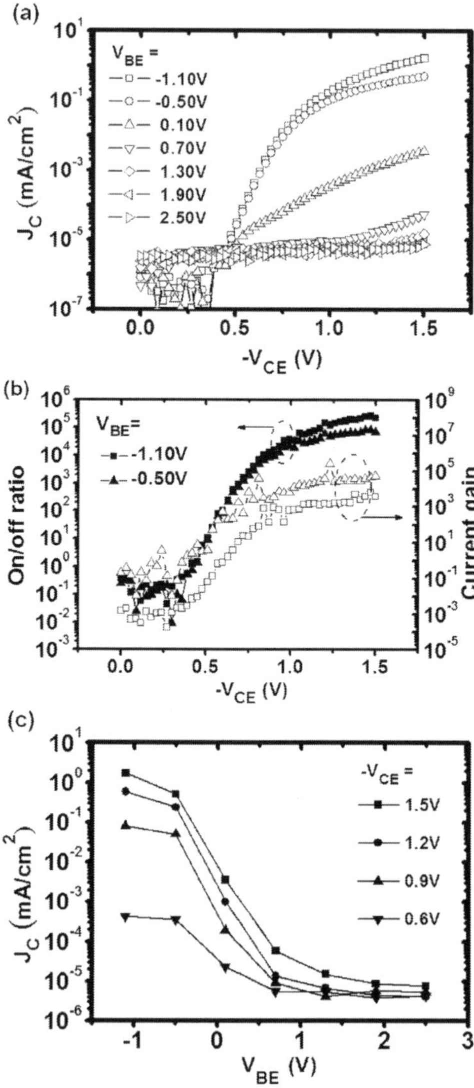

Figure 4.43 The variation of the current density and carrier mobility with increasing P3HT thickness.

of SCLT. Second, the ratio of emitter-to-base thickness, to the opening diameter is greater than that used in Section 4.3. Better grid control is obtainable with a higher ratio.

4.4.2 *The On Current of SCLT*

Flexible electronics based on solution-processed semiconductors is an exciting emerging technology with potential applications in very large areas. This technology enables displays and sensor arrays to cover the surfaces of clothes, buildings, robots, and cars [57, 58]. In such a large-area array, the pixel size may be over one centimeter and transistors need to drive high-power devices in the pixel as an organic light-emitting diode with high luminance or a motor and other mechanical actuator in robotic applications. Achieving a current output over 100 mA is difficult to achieve using polymer field-effect transistors. For a channel length of 50 μm typically made by a shadow mask, the drain current is only 4 mA for a channel width as long as 100 cm, given carrier mobility of 0.1 cm^2/Vs, drain bias of 20 V, and capacitance of 10 nF/cm^2. In principle, 50 pairs of a 1 cm long inter-digitized source and drain fingers can make a 100 cm channel width, resulting in a 1 cm by 1 cm active area, as shown in Fig. 4.44. Nevertheless, in addition to insufficient current output, fabricating this structure is fragile because any defect would break the entire finger. Furthermore, the operation voltage of an organic field-effect transistor is typically over 20 V, so the Joule heating and power consumption will be a major problem for high-current applications. The polymer field-effect transistor is therefore not suitable for driving a high-current load in large-area flexible arrays.

Organic vertical transistors offer another approach to this requirement [8–11, 25–34, 37]. This transistor has intrinsically low operation voltage and the current is directly proportional to its surface area for easy scaling up. One of the most successful vertical transistors is the SCLT, a solid-state version of the vacuum tube triode, with a metal grid embedded in an organic diode to control its vertical current [9–11, 33, 34]. The SCLT shows low operation voltage below 1 V and has a robust fabrication using self-assembled submicron plastic spheres as the shadow mask. However, so far its current density (J_C) is only 1 mA/cm^2 due to poor vertical mobility of the polymer. While the field-effect mobility of P3HT and some other polymers is over 0.1 cm^2/Vs, the vertical bulk mobility is only 10^{-5} cm^2/Vs. The surface treatment for the gate dielectric enhances chain alignment near the interface, but not in the bulk. High carrier density at the interface also increases mobility by filling the traps.

In this section, the polymer vertical mobility greatly enhances by a slow drying process using high-boiling-point solvent, i.e., solvent annealing, in a textured surface with a 200 nm scale defined by the grid. The vertical columns promote chain alignment of P3HT during slow solvent annealing, resulting in an ordered structure unattainable in a planar substrate. Vertical mobility raises from 4×10^{-5} cm^2/Vs to 2×10^{-3} cm^2/Vs and the SCLT can deliver output current density as high as 10 mA/cm^2 while maintaining good current gain and on–off ratio. With such high current density, a 1 cm^2 active area can deliver 100 mA current output, which is very easy to fabricate with high reliability. This grid-induced crystallinity SCLT solves the power transistor problem for large-area flexible electronics.

4.4.2.1 Fabrication process and device design

The SCLT was fabricated according to the process flow described in Section 4.3.1. In this section, the thickness of PVP is 210 nm. A 50-nm-thick SiO was evaporated sequentially after the base metal (a 40-nm-thick aluminum) evaporation to serve as the top insulator. 350 nm P3HT was then deposited with or without solvent annealing. For the device prepared by spin coating, P3HT in chlorobenzene

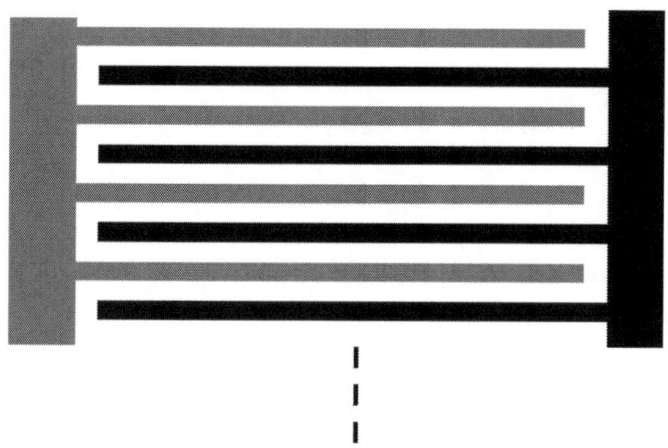

Figure 4.44 Interdigitized source and drain electrodes.

or dichlorobenzene was spin coated on the substrate until the film dried. As for the device prepared by dichlorobenzene solvent annealing, P3HT in dichlorobenzene (2.5 wt%) was spin coated at 450 rpm for 30 s and then 1200 rpm for 1.2 s to form a P3HT wet film. The wet film then remained still in the glove box until the film dried. The P3HT film was annealed at 140°C for 10 min. The active area was 1 mm^2. Transistor characteristics were recorded using an Agilent E5270B modular parameter analyzer. SEM image was obtained with a field-emission SEM (JEOL JSM-7401) The XRD pattern was obtained with a high resolution X-ray diffractometer, D1, Bede.

4.4.2.2 Characteristics of planar diodes and SCLT

Diodes and SCLTs with P3HT thickness of 350 nm were prepared from chlorobenzene (b.p. ~130°C) and dichlorobenzene (b.p. ~180°C) to understand the influence of the solvent and drying process on the device characteristics.

Because of the high boiling point of dichlorobenzene, the drying time was 7 s for spin coating for the P3HT chlorobenzene solution, while it was 4 min for spin coating the P3HT dichlorobenzene solution. Longer drying time of about 20 min was achieved by a solvent annealing procedure, which first spin coated the P3HT dichlorobenzene solution for a short time and then allowed the P3HT dichlorobenzene wet film to dry slowly in the glove box. The drying time was visually judged by the change in film color. Figure 4.45 shows the diode structure and characteristics. The output current density (J) of the P3HT diode prepared by spin coating chlorobenzene was higher than the one prepared by spin coating dichlorobenzene. With dichlorobenzene solvent annealing, J enhanced but was still inferior to the one achieved by chlorobenzene spin coating. The mobility, deduced from J following the SCLC of the diode prepared by spin coating P3HT chlorobenzene solution was therefore the highest one among various diodes, as shown in Fig. 4.46. The influence of the solvent and drying process on the SCLT was, however, not in the same trend as the diode shown in Fig. 4.48. Figure 4.47a,b show the SCLT structure. The active layer of SCLT was fabricated by spin coating P3HT chlorobenzene solution (SCLT-A), spin coating P3HT dichlorobenzene solution (SCLT-B), and

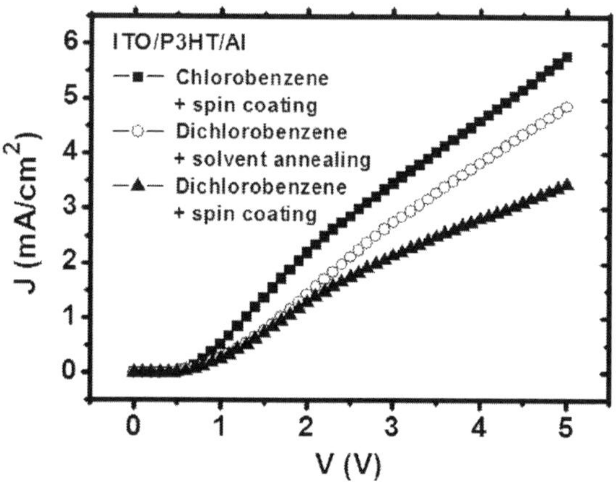

Figure 4.45 P3HT diode made by different method (P3HT thickness).

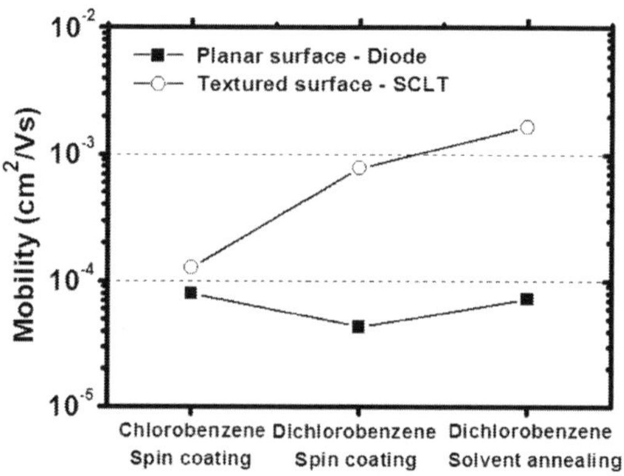

Figure 4.46 The influence of solvent annealing on carrier mobility.

spin coating P3HT dichlorobenzene solution for a time, and then allowing the wet film to dry slowly (SCLT-C). The ITO emitter was commonly grounded and the Al collector was negatively biased at collector-to-emitter voltage (V_{CE}). J_C was obtained by dividing the output current by device active area of 1 mm^2. The grid-to-emitter voltage (V_{GE}) was varied to modulate J_C. As the transistor turned on with $V_{GE} = -0.9$ V, J_C of SCLT-C (as shown in Fig. 4.48b) was higher than the one of SCLT-A (as shown in Fig. 4.48a) for over one

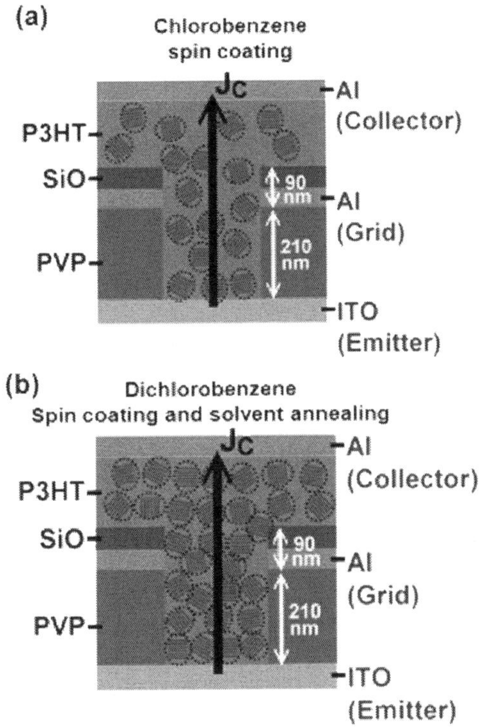

Figure 4.47 Device structures of SCLTs.

order of magnitude. Figure 4.48d shows that J_C is 36 mA/cm^2 for SCLT-B ($V_{GE} = -1.1$ V and $V_{CE} = -4.5$ V) while it is 52 mA/cm^2 for SCLT-C ($V_{GE} = -1$ V and $V_{CE} = -4.5$ V). J_C of the SCLT fabricated by dichlorobenzene was higher than the one fabricated by chlorobenzene. The SCLT fabricated with dichlorobenzene solvent annealing output a higher current than the one fabricated with spin coating. As a whole, high-boiling-point solvent and long drying time enhanced the output current. Furthermore, for SCLT-C, J_C of 104 mA/cm^2 was achieved when V_{CE} was -9.5 V, as shown in Fig. 4.48c. With operation voltage below 10 V, this transistor was able to deliver 100 mA output current by a 1 cm^2 area, which could drive a high-current load in large-area flexible arrays. As the SCLT was turned off, the off current of SCLT-C was still higher than SCLT-B and SCLT-A.

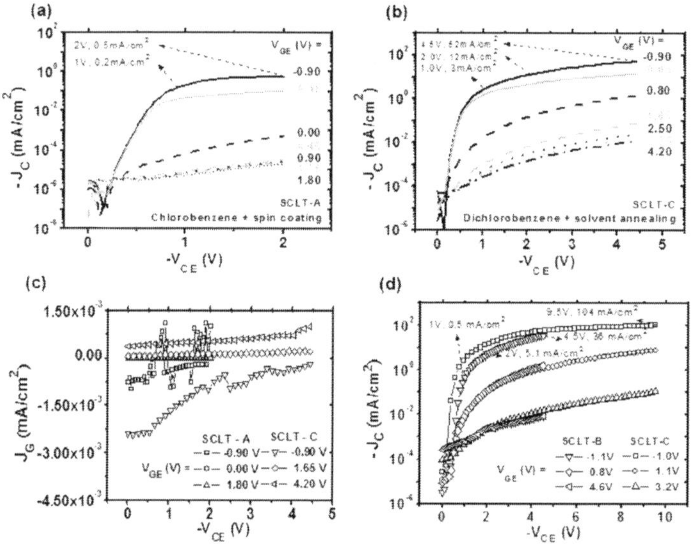

Figure 4.48 (a) The output characteristics of the SCLT made by spin coating chlorobenzene solution. (b) The output characteristics of the SCLT made by dichlorobenzene solvent annealing. (c) The grid current density. (d) The output characteristics of the SCLT operated at high voltage.

The output current of SCLT enhanced no matter whether it was in on state or off state. However, on/off ratios of these SCLTs were in the same order of 10^4. Figure 4.48c shows the grid current density (J_G) in the order of 10^{-4} to 10^{-3} mA/cm^2 and hence the current gain $| J_C/J_G |$ is as large as 10^4 to 10^5.

SEM images at difference phases of device fabrication explain why the on and off current of SCLT-C was higher than other devices. Figure 4.49a shows the top view of PS spheres on PVP. Figure 4.49b shows the openings forming at the sites without the Al grid. Although most PS spheres are separate from each other, aggregates containing more than two spheres are still observable. Large openings occur at sites with these randomly occurred aggregates. The grid control of the current is strong near the edge of the grid. The grid cannot effectively modulate the vertical carrier channel in the region far from the grid. As large openings occur, the grid control on the current far from the grid is small. Despite that the on current

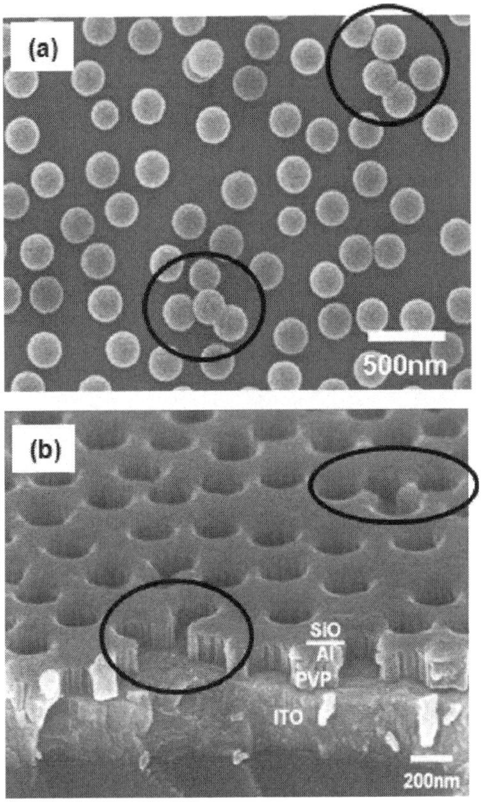

Figure 4.49 SEM images. (a) The top view of polystyrene spheres on PVP. (b) The openings formed at the sites without the Al grid. (a) The output characteristics of the SCLT made by spin coating chlorobenzene solution. (b) The output characteristics of the SCLT made by dichlorobenzene solvent annealing. (c) The grid current density. (d) The output characteristics of the SCLT operated at high voltage.

benefit from the slow drying process enhances carrier mobility, the V_{GE} cannot suppress the off current since the grid control on the current in the region away from the grid is less. The off current of SCLT is higher than J_G.

Accumulating PS spheres causes a significant increase of hole diameters. Figure 4.49 shows that when one PS sphere has a diameter of 200 nm, the accumulation of three PS spheres produce a hole-diameter of 200–600 nm. A large hole requires a large base potential to turn off the channel. A large base potential, however, breaks down the EC diode or EB diode, making it important to avoid accumulated PS spheres. Reducing the concentration of PS spheres to avoid accumulation can solve this problem. However, this method significantly reduces the on current because of the reduced effective channel, cross-sectional area per unit active area. Future development can use an ordered nano-hole structure to replace the current process. For example, nano imprint [59] and interference lithography [60] are low-cost promising technologies to make a metal grid with ordered nanometer holes to serve as the base electrode in SCLT.

4.4.2.3 Grid enhanced polymer crystallinity

Figure 4.46 shows the SCLT mobilities. Since the current channel forms only at the openings, this study recalculates the device active area according to the amount of openings for estimating current density and mobility. Mobilities deduced form SCLTs are higher than those from diodes except the one of SCLT-A. The highest mobility attainable in SCLT-C is 2×10^{-3} cm^2/Vs. With a high boiling point solvent and solvent annealing process, P3HT can self-organize into semicrystalline lamellar and yield high carrier mobility [61].

Carrier mobility is different in the diode and SCLT, explained as vertical channel columns inducing chain alignment of P3HT during slow solvent annealing. To understand why mobility increases in SCLT. Figure 4.50a–c shows atomic force microscopy (AFM) images of the P3HT surface on the textured surface. The chlorobenzene solution achieves root mean square roughness of 13.6 nm for spin-coated film. The roughness increases to 15.3 nm when using a high boiling point dichlorobenzene. A dichlorobenzene solvent annealing process further increases roughness to 22.2 nm. Research has demonstrated that polymer film with high surface roughness indicates polymer organization [62]. Increased SCLT mobility may come from enhanced polymer chain alignment. To confirm this, the current study uses grazing incidence X-ray diffraction (GIXRD), UV-visible absorption spectroscopy, and photoluminescence

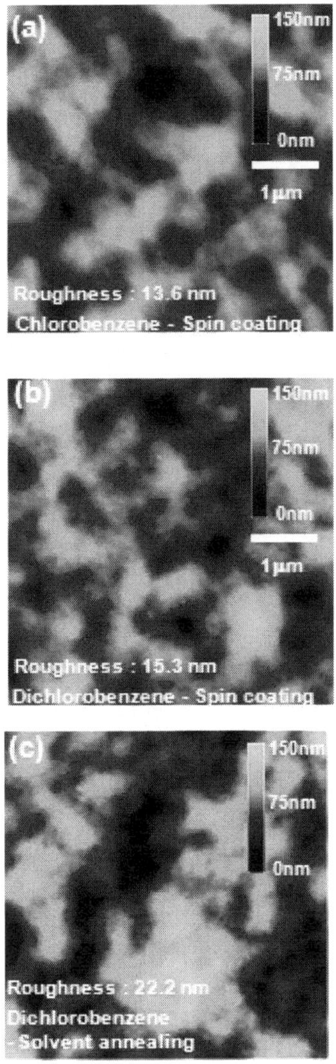

Figure 4.50 AFM images of P3HT surfaces fabricated by different method.

spectroscopy to obtain more in-depth information inside the film since AFM images only represent the film surface condition.

The out-of plane GIXRD spectra shown in Fig. 4.51 indicates stronger reflection of the (100) layer at $2\theta = 5.4°$ for film prepared with dichlorobenzene solvent annealing [63]. All P3HT films show

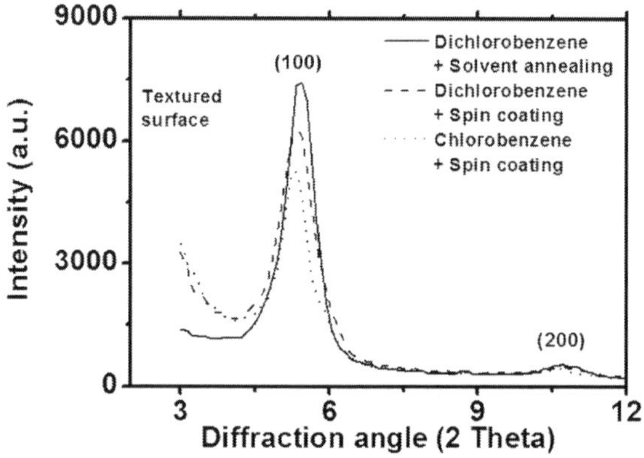

Figure 4.51 The out-of plane GIXRD spectra of P3HT surfaces fabricated by different method.

a well-organized intra plane structure, which means P3HT self-organizes into two-dimensional π-stacked lamellar structures. The intensity increases when using dichlorobenzene and the solvent annealing process. Crystallinity of P3HT improves using high boiling point dichlorobenzene and a solvent annealing process as illustrated in Fig. 4.47a,b.

The UV-visible absorption spectra in Fig. 4.52 shows prominent vibronic features at 600 nm for film prepared from dichlorobenzene. A previous work demonstrated that the enhanced vibronic absorption peak indicates strong interchain interactions and enhanced P3HT ordering [64]. This represents that using high boiling point dichlorobenzene helps induce ordered film with strong inter chain interactions. Figure 4.53a shows the influences of π-stacked polymer chains on photoluminescence spectra. The 0–0 to 0–1 peak intensity ratio, S_R, is typically used as an effective probe for disorder [65, 66] The 0–0 peak is forbidden by symmetry in disorder-free aggregates. Disorder breaks the symmetry and allows 0–0 emission. The S_R as high as P3HT is spin coated from chlorobenzene; however, using dichlorobenzene as the solvent decreases the S_R, reaching the lowest value when the solvent annealing process is used. This indicates that, on the textured surface, the P3HT film prepared from dichlorobenzene solvent annealing has less disorder, while the one prepared from chlorobenzene spin coating have

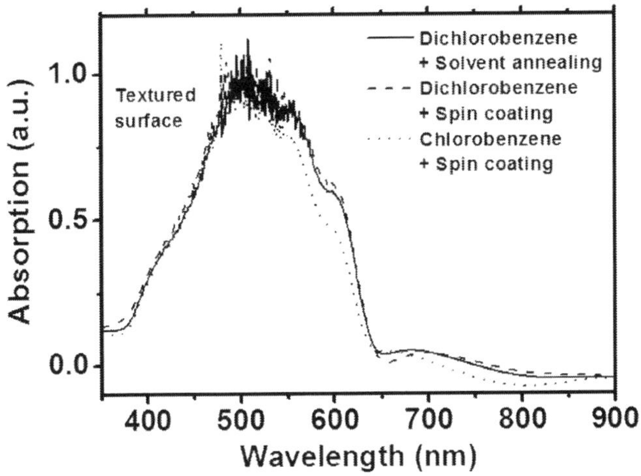

Figure 4.52 Absorption spectra of the P3HT surfaces fabricated by different method.

much disorder. Mobility is therefore high in P3HT film prepared from dichlorobenzene solvent annealing. A longer drying time may attribute to the well-ordered lamellar structure. However, on the planar surface, the film prepared from dichlorobenzene solvent annealing is most disorderly, while the film prepared from chlorobenzene spin coating is less disorderly, as shown in Fig. 4.53b. The difference in photoluminescence spectra of P3HT on the textured surface and the planar surface reveal the level of disorder and might explain the difference in device performance.

With this grid-induced crystallinity, SCLT treated with dichlorobenzene solvent annealing can output a high current suitable for large-area flexible electronics. Improved π–π stacking due to nanoconfinement induced chain alignment can explain improved P3HT ordering in the grid structure. Several other groups also reported nanostructure-induced crystallinity in P3HT [63, 67, 68]. K. M. Coakley *et al.* reported that P3HT nanopillars with diameters ranging from 20 to 75 nm have larger hole-mobility than P3HT neat film. For pore diameter above 75 nm hole-mobility falls off and approaches mobility value for a neat film [67]. In this experiment, with a pore diameter of 200 nm, hole mobilities in P3HT nanopillars were still higher than those in the neat film. The discrepancy may

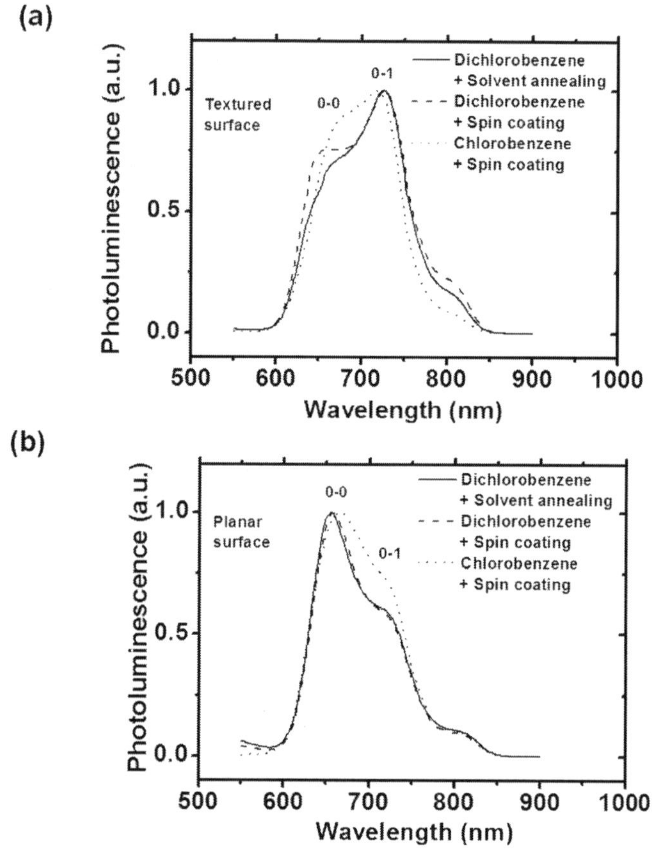

Figure 4.53 PL spectra of the P3HT on (a) textured surface and (b) planar surface.

come from the influences of solvent, drying method, and pore sidewall properties. Nanostructure-induced crystallinity, however, is a clear phenomenon in all these works.

4.5 Grid-Field Shielding in Polymer SCLTs

In SCLT, the potential in openings is controlled by the base grid electric field. Particularly, the channel is turned off when the grid field produces potential barrier to impede the transport of holes

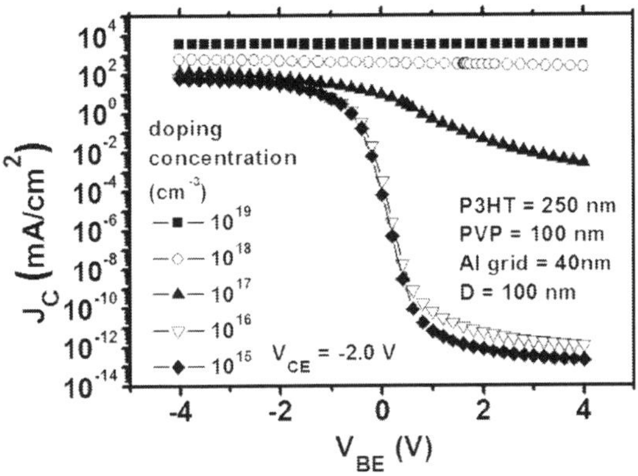

Figure 4.54 The influence of doping on the transfer characteristics; *L* is 250 nm, *D* is 100 nm, and PVP is 100 nm.

from emitter to collector. To successfully operate the transistor, the grid field penetrating through the organic semiconductor material into the center of openings is required. When dopants are added into the organic semiconductor, the high concentration holes, together with the ionized dopants, shield the grid field through the charge-shielding effect. As shown by the simulated J_C–V_{BE} curves in Fig. 4.54, when doping concentration is increases, the base control over channel potential profile is suppressed. In the following section, the grid-field shielding effect is investigated by adding dopants into channel material or by irradiating light to generate electron-hole pairs.

4.5.1 *Doping Effect*

To demonstrate the influence of controllable doping on transistor characteristics, this research uses the dopant F4-TCNQ, commonly used for controlled *p*-doping of small molecule organic materials and polymers, to dope P3HT and TFB. *p*-doping might occur via electron transfer from the HOMO of the polymer to the lowest unoccupied molecular orbital (LUMO) of the dopant. Doping not only increases bulk conductivity, but also enhances hole injection. For the SCLT made by F4-TCNQ-doped TFB, Fig. 4.55a shows the J–V curves of the emitter-collector diodes without applying

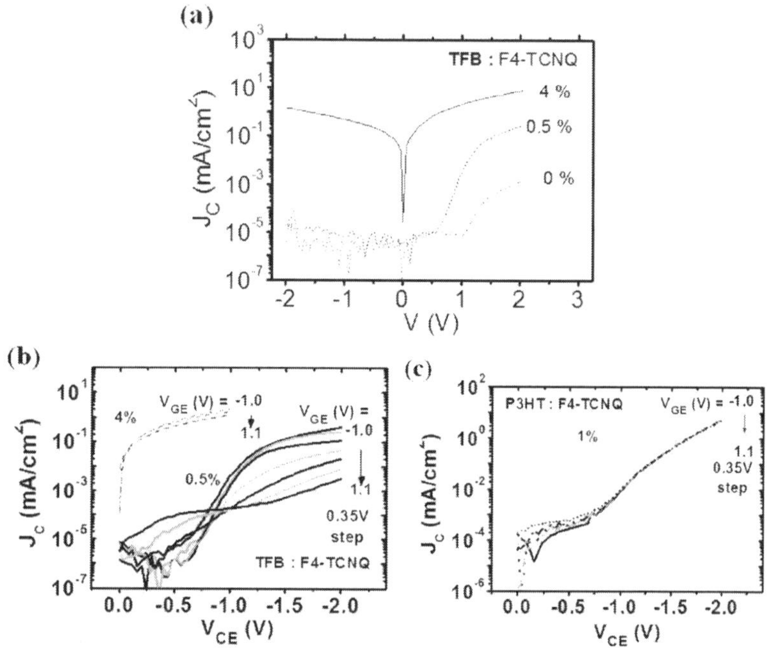

Figure 4.55 (a) The $J-V$ curves of the emitter-collector diodes without applying voltage on the grid. (b) The characteristics of SCLT made by F4-TCNQ-doped TFB. (c) The characteristics of SCLT made by F4-TCNQ doped P3HT [69].

voltage on the grid. With increasing doping concentration, the turn-on current increases and the turn-on voltage decreases. The results indicate that F4-TCNQ effectively increases conductivity in TFB and lowers the injection barrier as in previous reports [70]. Figure 4.55b shows the characteristics of SCLT made by F4-TCNQ-doped TFB. With 0.5% F4-TCNQ doping concentration, TFB SCLT exhibits an on/off ratio at about 120 under a 2 V VCE. However, when F4-TCNQ doping concentration increases to 4%, the on/off ratio turns to a low value, regardless of the high output current. Figure 4.55c shows a similar result obtained for the SCLT made by F4-TCNQ doped P3HT in a low dopant concentration (1%). The grid loses control of the potential barrier between the emitter and the collector with high dopant concentration. The following explains this behavior. A sufficiently high dopant concentration increases the doping level and decreases the Debye screening length. The electric field provided by the grid voltage is screened and hence the current control is lost. In SCLT and other similar vertical transistors, grid control is essential to turn off the channel. Even though doping can effectively enhance conductivity and hole-injection, the dopant screening effect seriously blocks the grid control and degrades the device performance.

4.5.2 *Illumination Effect*

Figure 4.56 shows the transfer characteristics (i.e., the collector current density, J_C, plotted as a function of the base-to-emitter voltage, V_{BE}) of P3HT SCLT in the dark and under illumination when $V_{CE} = -1.5$ V In the dark, the off current is lower than 5×10^{-4} mA/cm^2. Under illumination, the off current increases significantly. If defining photo current density (J_{ph}) as the current density under illumination minus the current density in the dark (J_{dark}), J_{ph} as 1 mA/cm^2 is obtained when illumination intensity is 11 mW/cm^2, V_{BE} = 1.5 V, and V_{CE} = -1.5 V. The high J_{ph} in P3HT SCLT does not form by the photogenerated carriers in P3HT because

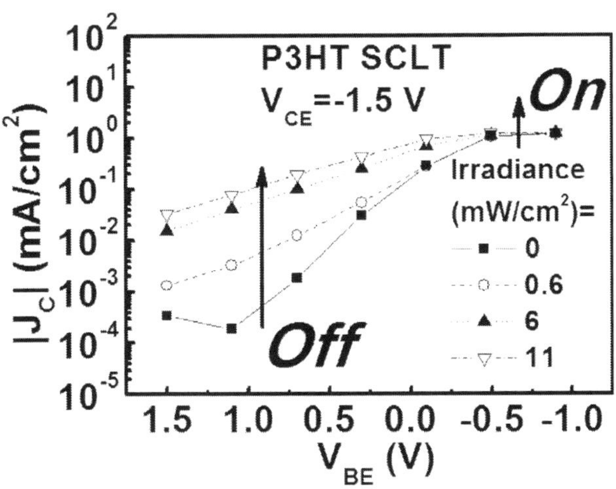

Figure 4.56 Transfer characteristics of a P3HT-SCLT in total darkness and under illumination [69].

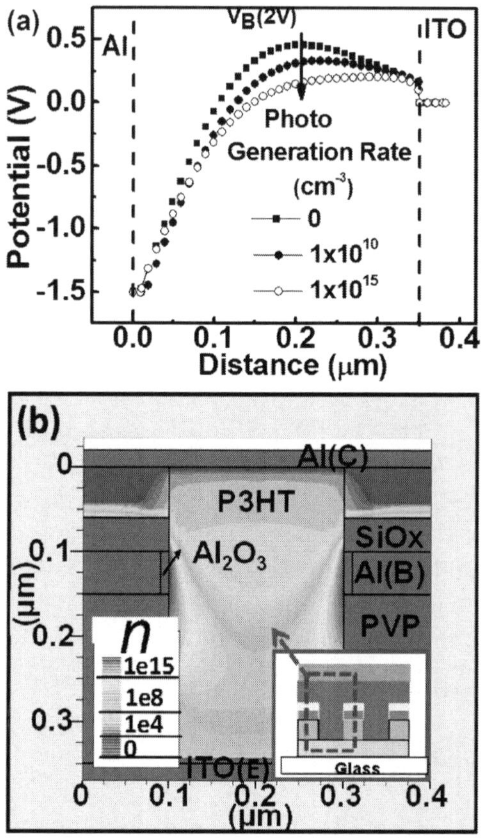

Figure 4.57 (a) A simulated potential distribution at the central vertical channel. (b) The two-dimensional electron distribution of P3HT-SCLT with a photo generation rate = 1010 cm^{-3} and a $V_{BE} = 2$ V. Electron concentration is denoted by n [69]. See also Color Insert.

the exciton dissociation in P3HT without an acceptor blending is poor. Poor exciton dissociation confirms by measuring the J_C–V_{CE} curve in a reverse-biased EC diode with a floating base potential in total darkness and under illumination (not shown). This study obtained low J_{ph} of 3×10^{-4} mA/cm^2 with illumination intensity of 11 mW/cm^2 and $V_{CE} = 1.5$ V.

To investigate the mechanism causing high J_{ph} in P3HT SCLT, simulated carrier distribution and the potential channel profile with

TCAD Silvaco Atlas software are performed. The current research performed a two-dimensional simulation by setting $V_{CE} = -1.5$ V and $V_{BE} = 2$ V to simulate the device operated in its off state. The definition of material parameters was according to Ref. [71]. Figure 4.57a shows the potential distribution in the central channel from emitter to collector. In total darkness (i.e., photo generation rate $= 0$ cm^{-3}), the emitter was unable to inject holes into the channel because the 2 V V_{BE} created a potential barrier to impede hole transport, achieving a low off-state current. Under illumination (i.e. photo generation rate $= 10^{10}$ or 10^{15} cm^{-3}), the potential barrier formed by the 2 V V_{BE} significantly dropped, and with a lower potential barrier, high off-state current was expected. The two-dimensional electron distribution plot shown in Fig. 4.57b explains the lower potential barrier under illumination. With a photo generation rate $= 10^{10}$ cm^{-3} and a $V_{BE} = 2$ V, the electrons generated by exciton dissociation accumulate around the electrode base. The accumulated electrons shield the base field and suppress the base bias influence on channel potential. As a result, the potential barrier caused by base bias lowered under illumination. The off-state current, with exponential dependence on potential barrier height, increased significantly.

However, due to the poor exciton dissociation in P3HT, electron concentration around the base electrode was not high enough to shield the grid-field completely. Accordingly, in Fig. 4.56, the off-state current density under irradiance as 11 mW/cm^2 is still 40 times lower than the on-state current. Under irradiance as 11 mW/cm^2 is still 38 times lower than the on-state current density. If further enhancement of the photoresponse expects complete grid-field shielding, the exciton dissociation in the channel region needs to be improved. The following experiment added PCBM to P3HT to form a donor/acceptor interface to improve exciton dissociation.

4.5.3 *Polymer Vertical Phototransistor*

Organic photodetector arrays can fabricate at low temperatures on a large scale, with low production costs. Integrating organic photodetectors and other organic electronic devices such as organic transistors, organic light-emitting diodes, and various sensors enables the development of numerous devices, including flexible touch panel displays, flexible large-area scanners, footprint scanners, proximity sensors, and biomedical sensors.

A good photodetector requires high photoresponsivity, low operational voltage, wide bandwidth, and processes that are compatible with organic transistors, for an active-matrix photodetection array. The term photodetectors generally refers to two classes of organic devices. The first class refers to an organic photodiode (OPD) with a vertical metal/semiconducting/metal structure. The semiconducting layer is generally a blend of a donor material with an acceptor material to absorb light and dissociate excitons. OPDs typically operate at low voltages (<5 V) with wide bandwidths (over 80 MHz) [72, 73]. Unfortunately, the processes involved in fabricating vertical structures are incompatible with the process of forming conventional horizontal organic thin-film transistor (OTFT). The second class of photodetector refers to organic phototransistors (OPT) based on an OTFT structure. The photodetection mechanism is based on the shift of threshold voltage due to formation of a charged state in the organic semiconducting layer or at the semiconductor/dielectric interface under light irradiation [74]. OPTs exhibit high photoresponsivity (>10 A/W), but typically require high operating voltage (>10 V).

This section introduces a new vertical OPT based on SCLT. This SCLT-based OPT utilizes a new photo detecting mechanism referred to as grid-field shielding, to induce photomultiplication (PM). Several research teams have reported the effects of PM in organic semiconducting devices [75, 76]. In 2007, Jinsong Huang and Yang Yang suggested that PM was due to photo-induced carrier injection [77]. In SCLT-based OPT, a new PM effect was found to be not only related to photo-induced carrier injection, but also greatly enhanced by the effects of grid-field shielding. To enhance the effect of grid-field shielding, one can exchange the channel material from donor material, P3HT, to a blend of donor material and an acceptor material, P3HT blended with PCBM, to enhance the dissociation of excitons. The SCLT-based OPT and P3HT blended with PCBM channel material demonstrated high EQE (approximately 358%), high photo-to-dark current ratio (approximately 2000), low operational

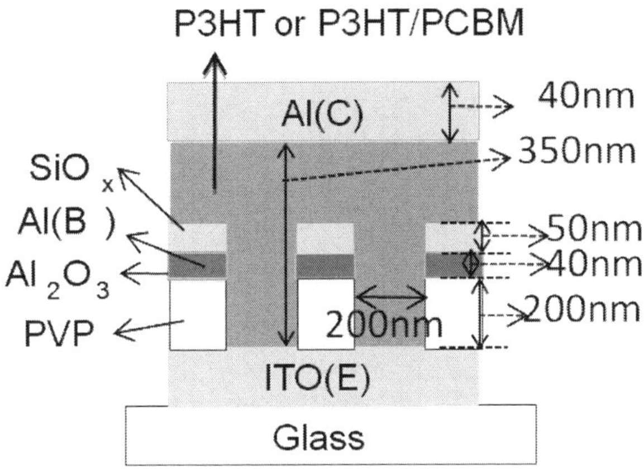

Figure 4.58 Schematic illustration of the structure of P3HT or P3HT/PCBM-SCLT [69].

voltage (−1.5 V), and positive transistor characteristics with an initial on/off current ratio as high as 5000 for single devices.

4.5.3.1 Vertical phototransistor fabrication process and basic characteristics

Figure 4.58 shows the device structure Section 4.3.1 describes the fabrication process. The thickness of PVP is 210 nm. Two kinds of active layer were used. One was P3HT of 3500 Å from 4.5 wt% chlorobenzene to form a conventional P3HT SCLT. The other was a blended P3HT/PCBM of 3500 Å from 4.5 wt% chlorobenzene to form a new P3HT/PCBM SCLT. The P3HT:PCBM blending ratio was 1:0.1 by weight.

Figure 4.59a shows the characteristics of the P3HT/PCBM SCLT. When base to emitter potential (V_{BE}) changed from −0.9 V to 1.5 V, the device was switched from on state to off state. Maximum on/off current ratio of approximately 5000 was obtained at a collector to emitter potential (V_{CE}) of −1.5 V. The on/off current ratio was inferior to that of P3HT SCLT because electron injection from the collector metal (aluminum) to PCBM caused a leakage in the current. Applying an electron-blocking layer between the collector and active layer could reduce the leakage in future studies. To maintain a

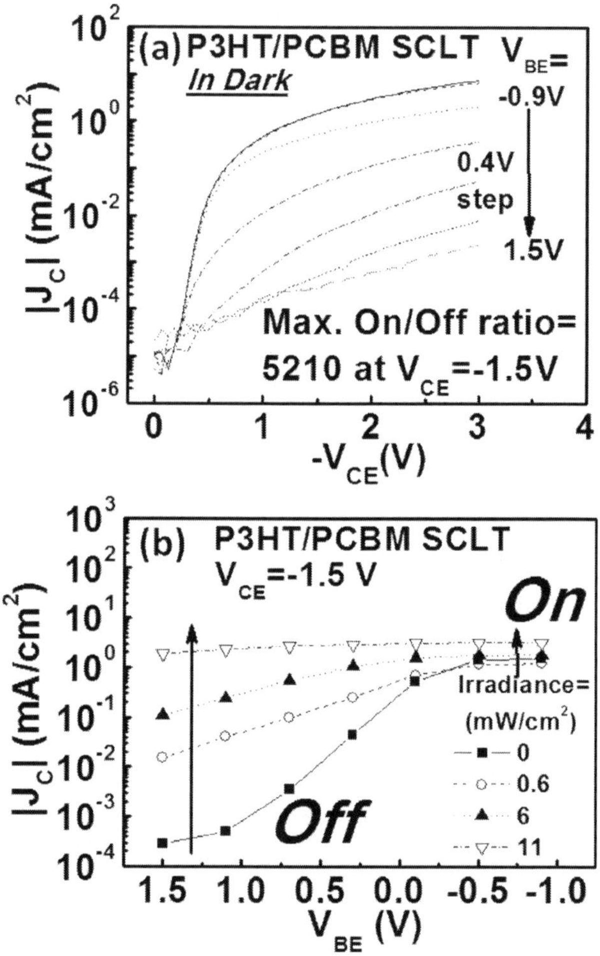

Figure 4.59 (a) Output characteristics of the P3HT/PCBM-SCLT. (b) Transfer characteristics of a P3HT/PCBM SCLT in total darkness and under illumination [69].

reasonable on/off current ratio, the P3HT/PCBM blending ratio is kept to be 1:0.1 by weight. If the P3HT/PCBM blending ratio was 1:0.5 or 1:1, the device would suffer from large leakage current, which would degrade switching performance.

4.5.3.2 Photocurrent analysis

In Fig. 4.59b, the *on* current increases significantly under illumination. Because the SCLT operated in on state is similar to a forward-biased diode, the following work analyzed the photoresponse of a P3HT/PCBM diode in both reverse and forward bias conditions. Two planar ITO/P3HT/PCBM/Al diodes are fabricated; one with a P3HT/PCBM blending ratio of 1:0.1 (Diode A), and the other with a P3HT/PCBM blending ratio of 1:1 (Diode B), and plotted the J_{ph} of these two diodes as a function of bias voltage in Fig. 4.60a with an illumination intensity of 11 mW/cm^2 In reverse-biased conditions, Diode A had a much lower J_{ph} than Diode B, because the low PCBM percentage suppressed exciton dissociation. In forward-biased conditions, J_{ph} of Diode A was 20 times higher than that of Diode B. For Diode B, J_{ph} in reverse-biased condition was similar to that in forward-biased condition, indicating that exciton dissociation dominated both cases. For Diode A, J_{ph} in forward-biased condition was 100 times greater than J_{ph} in reversed-biased condition. Obviously, exciton dissociation does not explain the large J_{ph} in forward-biased condition. It is proposed that illumination enhanced the hole-injection from ITO to P3HT. In previous reports, photo-induced carrier injection was a result of carrier trapping at the injection interface to lower the injection barrier [77]. This study proposes that electrons trapped at the ITO/channel interface enhanced hole injection. Figure 4.60a shows the J_{ph} of EC diode in P3HT/PCBM SCLT by a dotted line. A J_{ph} higher than 1 mA/cm^2 is obtained, when forward-biased voltage was 1.5 V (i.e., $V_{CE} = -1.5$ V). Although the forward-biased Diode A and the forward-biased EC diode had high J_{ph}, their photo-to-dark current ratios (J_{ph}/J_{dark}) were smaller than 1, because of the high J_{dark}. In P3HT/PCBM SCLT, the base potential in total darkness turned off the current flowing through a forward-biased EC diode (i.e., $V_{CE} = -1.5$ V), achieving a low J_{dark}. Under illumination, the base electrodes lost control over the channel current because of the effect of grid-field shielding. At the same time, the effects of photo-enhanced hole-injection in the forward-biased EC diode contributed to a high channel current. The photo-to-dark current ratios (J_{ph}/J_{dark}) were 0.12, 0.17, and 0.44, when illumination intensities were 0.61, 5.88, and 11 mW/cm^2.

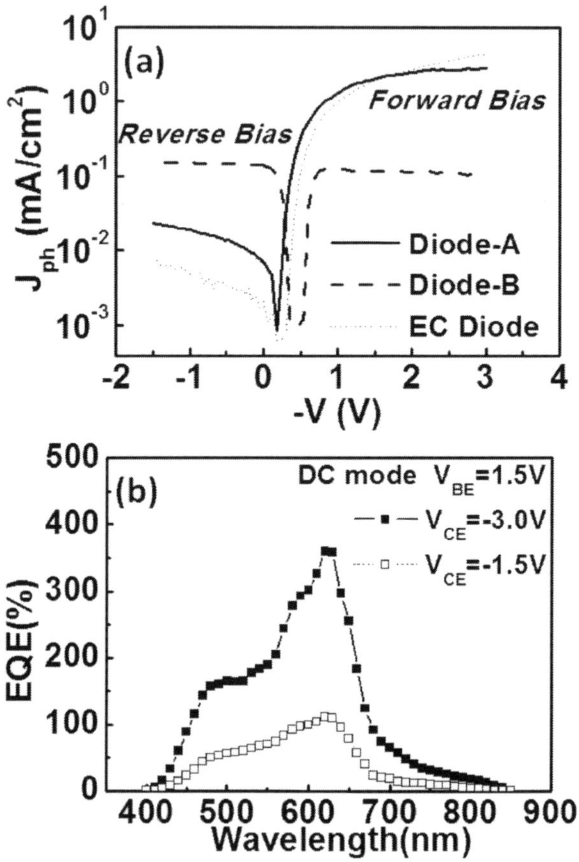

Figure 4.60 (a) The J_{ph} of Diode A, Diode B, and EC diode in P3HT/PCBM-SCLT plotted as a function of bias voltage. (b) The EQE of P3HT/PCBM-SCLT [69].

4.5.3.3 External quantum efficiency measurement and photomultiplication effect

Finally, the measured external quantum efficiency (EQE) of the P3HT/PCBM SCLT with an active area of 1 mm^2, $V_{\text{BE}} = 1.5$ V and $V_{\text{CE}} = -1.5$ or -3 V, is shown in Fig. 4.60b. In SCLT, when irradiated from an ITO electrode, only a fraction of the active area comprised P3HT/PCBM material. The EQE measured in this case was actually an effective EQE per unit active area. The maximum EQE of P3HT/PCBM SCLT was 110% and 358% at 620 nm when V_{CE} = −1.5 V or −3 V. Grid-field shielding effect dominated an EQE higher than 100, representing a PM phenomenon. The high EQE, however, had a narrow bandwidth. When the measured EQE frequency increased to 80 Hz and 160 Hz, maximum EQE dropped to 15 and 12. Slow electron accumulation surrounding the base electrode explains the narrow bandwidth. Under illumination, excitons dissociated at P3HT/PCBM interface to generate free electrons. Electrons moved from the P3HT/PCBM interface to the base electrode to accumulate around the base. However, low electron mobility and strong recombination of electron-hole pair in P3HT suppressed the rate of electron accumulation. The response of the grid-field shielding effect was therefore slow. In future studies, incorporating more PCBM into P3HT may succeed in enhancing the speed of grid-field shielding, while preventing leakage through PCBM would require an electron-blocking layer between the collector (aluminum) and the active layer.

The current study also measured the EQE of a reverse-biased EC diode in P3HT/PCBM SCLT (not shown). The maximum EQE was 0.3% and 1.4% at 620 nm when V_{CE} = 1.5 V and 3 V The EQE in the EC diode was much lower than the EQE for a normal P3HT/PCBM photodetector. A normal P3HT/PCBM photodetector, with an active layer thickness of 250 nm and a P3HT/PCBM blending ratio as 1:1, typically had an EQE of 60% at 600 nm, when the bias voltage was 3 V [78]. The following explains the low EQE in the EC diode of SCLT: (1) In SCLT, only a fraction of the active area comprised P3HT/PCBM material; (2) The P3HT/PCBM blending ratio was 1:0.1. Figure 4.60a shows that when diodes are reverse biased, J_{ph} of diode with P3HT/PCBM = 1:0.1 is one order lower than that of the diode with P3HT/PCBM = 1:1.

In summary, a vertical polymer phototransistor, based on the SCLT is demonstrated. The following steps governed photoresponse in SCLT. First, the SCLT operated in the off state to provide a current in total darkness. The EC diode was forward biased and the base potential created a potential barrier to impede hole-transport. Second, under illumination, excitons generated in the polymer channel layer. In this study, P3HT blended the channel material with PCBM at a blending ratio of 1:0.1, to enhance exciton dissociation.

Third, after exciton dissociation, electrons flowed toward the base electrode. Electrons accumulated around the base electrode shield of the base field, reducing the potential barrier. At the same time, a portion of the electrons trapped at the ITO/channel interface, enhanced the hole-injection from ITO into the channel. Finally, the reduced potential barrier and enhanced hole-injection obtained a large photocurrent.

4.6 Conclusion

In this chapter, we introduce the theoretical analysis and some recent progress of polymer vertical transistors. The polymer vertical transistor based on the concept of a solid-state vacuum tube, which is the polymer vertical transistor with the highest on–off current ratio at a low operation voltage, is particularly addressed. The revolution of the polymer vertical transistor includes the reduction of the operation voltage, the suppression of the leakage current, the improvement of the base grid control over channel, and the improvement of output current by increasing the vertical mobility, etc. Now state-of-art polymer vertical transistors show enough performances to be applied on some low-end applications. To develop a roll-to-roll manufacturing process, ink-jet printing, blade coating, or screen printing are candidates to serve as the process platform. Nano-imprint is also promising to produce the metal grid. A low-cost and power-saving application using polymer vertical transistor will be expected in near future.

References

1. W. Shockley (1952). *Proc. IRE* **40**, 1289.
2. S. A. Roosild, R. P. Dolan, and D. O'Neil (1963). *Proc. IEEE* **51**, 1059.
3. B. Buchanan, S. Roosild, and R. Dolan (1964). *Proc. IEEE* **52**, 1364.
4. S. Teszner and R. Gicquel (1964). *Proc. IEEE* **52**, 1502.
5. R. Zuleeg and V. O. Hinkle (1964). *Proc. IEEE* **52**, 1245.
6. K. Kudo, D. X. Wang, M. Iizuka, S. Kuniyoshi, and K. Tanaka (1998). *Thin Solid Films* **331**, 51.

7. Y. Yang and A. J. Heeger (1994). *Nature* **372**, 344.
8. K. Fujimoto, T. Hiroi, K. Kudo, and M. Nakamura (2007). *Adv. Mater.* **19**, 525.
9. Y. C. Chao, M. C. Niu, H. W. Zan, H. F. Meng, and M. C. Ku (2011). *Org. Electron.* **12,** 78.
10. Y. C. Chao, M. C. Ku, W. W. Tsai, H. W. Zan, H. F. Meng, H. K. Tsai, and S. F. Horng (2010). *Appl. Phys. Lett.* **97**, 223307.
11. Y. C. Chao, Y. C. Lin, M. Z. Dai, H. W. Zan, and H. F. Meng (2009). *Appl. Phys. Lett.* **95**, 203305.
12. D. K. Hwang, K. Lee, J. H. Kim, S. Im, J. Hoon, and E. Kim (2006). *Appl. Phys. Lett.* **89**, 093507.
13. S. D. Wang, T. Minari, T. Miyadera, K. Tsukagoshi, and Y. Aoyagi (2007). *Appl. Phys. Lett.* **91**, 203508.
14. S. H. Kim, J. Jang, H. Jeon, W. M. Yun, S. Nam, and C. E. Park (2008). *Appl. Phys. Lett.* **92**, 183306.
15. C. H. Wang, S. W. Chen, and J. Hwang (2009). *Appl. Phys. Lett.* **95**, 103302.
16. Y. Hu, Q. Qi, and C. Jiang (2010). *Appl. Phys. Lett.* **96**, 133311.
17. M. L. Chabinyc, F. Endicott, B. D. Vogt, D. M. DeLongchamp, E. K. Lin, Y. Wu, P. Liu, and B. S. Ong (2006). *Appl. Phys. Lett.* **88**, 113514.
18. C. R. Newman, H. Sirringhaus, J. C. Blakesley, and R. Speller (2007). *Appl. Phys. Lett.* **91**, 142105.
19. S. Wang, J. C. Tang, L. H. Zhao, R. Q. Png, L. Y. Wong, P. J. Chia, H. S. O. Chan, P. K.-H. Ho, and L. L. Chua (2008). *Appl. Phys. Lett.* **93**, 162103.
20. Y. M. Park and A. Salleo (2009). *Appl. Phys. Lett.* **95**, 133307.
21. M. Caironi, C. Newman, J. R. Moore, D. Natali, H. Yan, A. Facchetti, and H. Sirringhaus (2010). *Appl. Phys. Lett.* **96**, 183303.
22. M. Zirkl, A. Haase, A. Fian, H. Schön C. Sommer, G. Jakopic, G. Leising, B. Stadlober, I. Graz, N. Gaar, R. Schwödiauer, S. Bauer-Gogonea, and S. Bauer (2007). *Adv. Mater.* **19**, 2241.
23. Y. Lu, W. H. Lee, H. S. Lee, Y. Jang, and K. Cho (2009). *Appl. Phys. Lett.* **94**, 113303.
24. O. Acton, G. G. Ting, P. J. Shamberger, F. S. Ohuchi, H. Ma, and A. K.-Y. Jen (2010). *ACS Appl. Mater. Interfaces.* **2**, 511.
25. C. Y. Yang, T. M. Ou, S. S. Cheng, M. C. Wu, S. Y. Lin, I. M. Chan, and Y. J. Chan (2006). *Appl. Phys. Lett.* **89**, 183511.
26. T. M. Ou, S. S. Cheng, C. Y. Huang, M. C. Wu, I. M. Chan, S. Y. Lin, and Y. J. Chan (2006). *Appl. Phys. Lett.* **89**, 183508.

27. Y. C. Chao, M. H. Xie, M. Z. Dai, H. F. Meng, S. F. Horng, and C. S. Hsu (2008). *Appl. Phys. Lett.* **92**, 093310.
28. S. S. Cheng, Y. C. Chuang, K. Dhananjay, C. W. Ou, M. C. Wu, and C. W. Chu (2009). *Adv. Mater.* **21**, 1860.
29. S. S. Cheng, G. Y. Chen, J. H. Chen, M. C. Wu, and C. W. Chu (2010). *Org. Electron.* **11**, 692 .
30. Y. Watanabe, H. Iechi, and K. Kudo (2008). *Thin Solid Film* **516**, 2731.
31. F. Pu, Y. Watanabe, H. Yamauchi, M. Nakamura, and K. Kudo (2009). *Thin Solid Film* **518**, 514 .
32. H. Iechi, Y. Watanabe, H. Yamauchi, and K. Kudo (2010). *Jpn. J. Appl. Phys.* **49**, 01AB12.
33. Y. C. Chao, H. F. Meng, and S. F. Horng (2006). *Appl. Phys. Lett.* **88**, 223510.
34. Y. C. Chao, H. F. Meng, S. F. Horng, and C. S. Hsu (2008). *Org. Electron.* **9**, 310.
35. T. Someya, Y. Kato, T. Sekitani, S. Iba, Y. Noguchi, Y. Murase, H. Kawaguchi, and T. Sakurai (2005). *Proc. Natl. Acad. Sci. USA* **102**, 12321.
36. K. N. Kwok (2002). *Complete Guide to Semiconductor Devices*, 2nd ed., Wiley, New York .
37. Y. C. Chao, S. L. Yang, H. F. Meng, and S. F. Horng (2005). *Appl. Phys. Lett.* **87**, 253508.
38. L. Ma and Y. Yang (2004). *Appl. Phys. Lett.* **85**, 5084.
39. S. H. Li, Z. Xu, G. Yang, L. Ma, and Y. Yang (2008). *Appl. Phys. Lett.* **93**, 213301.
40. S. H. Li, Z. Xu, L. Ma, C. W. Chu, and Y. Yang (2007). *Appl. Phys. Lett.* **91**, 083507.
41. Y. Yang and A. J. Heeger (1994). *Nature* **372**, 344.
42. K. Kudo, M. Iizuka, S. Kuniyoshi, and K. Tanaka (2001). *Thin Solid Films* **393**, 362.
43. K. Fujimoto, T. Hiroi, and M. Nakamura (2005). *e-J. Surf. Sci. Nanotechnol.* **3**, 327.
44. N. Muraishi, M. Nakamura, M. Iizuka, and K. Kudo (August 2003), presented at *ICEL-4, 4th Int. Conf. on Electroluminescence of Molecular Materials and Related Phenomena*, Jeju, Korea.
45. N. Ohashi, M. Nakamura, N. Muraishi, M. Sakai, and K. Kudo (2006). *IEICE Trans. Electron.* **E89-C**, 1765.
46. N. Hirashima, N. Ohashi, M. Nakamura, and K. Kudo (2005). Proc. *Int. Symp. on Super- Functionality Organic Devices, IPAP Conference Series 6, The Institute of Pure and Applied Physics*, Tokyo, Japan, 158.

47. D. J. Milliron, I. G. Hill, C. Shen, A. Kahn, and J. Schwartz (2000). *J. Appl. Phys.* **87**, 572.
48. K. Sugiyama, H. Ishii, Y. Ouchi, and K. Seki (2000). *J. Appl. Phys.* **87**, 295.
49. K. H. Lee, H. W. Jang, K. B. Kim, Y. H. Tak, and J. L. Lee (2004). *J. Appl. Phys.* **95**, 586.
50. J. Y. Deng, Z. Z. You, W. Tao, L. W. Zhi, and J. X. Qiao (2006). *Proc. of SPIE* **6030**, 60300G.
51. Z. Chiguvare and V. Dyakonov (2004). *Phys. Rev. B* **70**, 235207.
52. K. D. Kim and C. K. Song (2006). *Appl. Phys. Lett.* **88,** 233508.
53. H. Kang, K. K. Han, J. E. Park, and H. H. Lee (2007). *Org. Electron.* **8,** 460.
54. Y. C. Chao, C. Y. Chen, H. W. Zan, and H. F. Meng (2010). *J. Phys. D: Appl. Phys.* **43**, 205101.
55. T. Y. Chu and O. K. Song (2007). *Appl. Phys. Lett.* **91**, 073508.
56. T. Y. Chu and O. K. Song (2008). *J. Appl. Phys.* **104**, 023711.
57. T. Someya, T. Sekitani, S. Iba, Y. Kato, H. Kawaguchi, and T. Sakurai (2004). *Proc. Natl. Acad. Sci. USA* **101**, 9966.
58. T. Sekitani, H. Nakajima, H. Maeda, T. Fukushima, T. Aida, K. Hata, and T. Someya (2009). *Nat. Mater.* **8** 494.
59. L. Jay Guo (2004). *J. Phys. D: Appl. Phys.* **37** R123.
60. C. Lu and R.H. Lipson (2010). *Laser Photonics Rev.* **4**, 568.
61. G. Li, Y. Yao, H. Yang, V. Shrotriya, G. Yang, and Y. Yang (2007). *Adv. Funct. Mater.* **17**, 1636.
62. G. Li, V. Shrotriya, J. Huang, Y. Yao, T.Moriarty, and K. Emery (2005). *Y. Yang, Nat. Mater.* **4**, 864.
63. J. S. Kim, Y. Park, D. Y. Lee, J. H. Lee, J. H, Park, J. K. Kim, and K. Cho (2010). *Adv. Funct. Mater.* **20**, 540.
64. P. J. Brown, D. S. Thomas, A. Kohler, J. S. Wilson, J. S. Kim, C. M. Ramsdale, H. Sirringhaus, and R. H. Friend (2003). *Phys. Rev. B* **67**, 064203.
65. J. Clark, C. Silva, R. H. Friend, and F. C. Spano (2007). *Phys. Rev. Lett.* **98**, 206406.
66. F. C. Spano, J. Clark, C. Silva, and R. H. Friend (2009). *J. Chem. Phys.* **130**, 074904.
67. K. M. Coakley, B. S. Srinivasan, J. M. Ziebarth, C. Goh, Y. Liu, and M. D. McGehee (2005). *Adv. Funct. Mater.* **15**, 1927.
68. M. Aryal, K. Trivedi, and W. Hu (2009). *ACSNano* **3**, 3085.
69. H. W. Zan, W. W. Tsai, and H. F. Meng (2011). *Appl. Phys. Lett.* **98**, 053305.

70. K. H. Yim, G. L. Whiting, C. E. Murphy, J. J. M. Halls, J. H. Burroughes, R. H. Friend, and J. S. Kim (2008). *Adv. Mater.* **20**, 3319.
71. The thickness of PVP, Al grid, SiO_x, and P3HT are 200, 40, 50 and 350 nm, respectively. The opening diameter is 200 nm. The highest occupied molecular orbital and lowest unoccupied molecular orbital levels of P3HT are 5.2 and 3.0 eV. The work function of emitter and collector are 5.2 and 4.3 eV. The hole mobility and electron mobility in P3HT are 10^{-5} and 10^{-6} cm^2/Vs.
72. P. Peumans, V. Bulovic, and S. R. Forrest (2000). *Appl. Phys. Lett.* **76**, 3855.
73. W. W. Tsai, Y. C. Chao, E. C. Chen, H. W. Zan, H. F. Meng, and C. S. Hsu (2009). *Appl. Phys. Lett.* **95**, 213308.
74. M. Debucquoy, S. Verlaak, S. Steudel, K. Myny, J. Genoe, and P. Heremans (2007). *Appl. Phys. Lett.* **91**, 10.
75. J. Reynaert, V. I. Arkhipov, P. Heremans, and J. Poortmans (2006). *Adv. Funct. Mater.* **16**, 784.
76. M. Hiramoto, T. Imahigashi, and M. Yokoyama (1994). *Appl. Phys. Lett.* **64**, 187.
77. J. Huang and Y. Yang (2007). *Appl. Phys. Lett.* **91**, 203505.
78. E. C. Chen, C. Y. Chang, J. T. Shieh, S. R. Tseng, H. F. Meng, C. S. Hsu, and S. F. Horng (2010). *Appl. Phys. Lett.* **96**, 043507.

Chapter 5

Polymer Solar Cells

Shang-Chieh Chien[a] and Fang-Chung Chen[b]
[a]*Department of Photonics and Institute of Electro-Optical Engineering, and*
[b]*Department of Photonics and Display Institute,*
National Chiao Tung University, Hsinchu, Taiwan
fcchen@mail.nctu.edu.tw

Organic solar cells, particularly those made using conjugated polymers and fullerenes, have attracted much attention because of their low-cost fabrication, mechanical flexibility, and light weight. In this chapter, we describe the evolution and fundamental properties of organic solar cells and then provide a review of recent developments in high-performance polymer solar cells. The power conversion efficiencies of polymer solar cells are generally improved through modifications of the electrode contacts. Improving the optical properties of polymer solar cells is an alternative approach toward enhancing device performance—for example, by using metal nanoparticles to induce localized surface plasmons, nanostructures to trap light, and optical spacers to redistribute the electric field in devices. Finally, several novel device architectures, including inverted, stacked, and tandem structures are introduced. Particularly high air-stability can be achieved in inverted structures. Stacked and tandem cells can provide higher photovoltages and enlarge the spectral coverage, allowing the collection of more photons from the sun.

Polymer Electronics
Edited by Hsin-Fei Meng

ISBN 978-981-4267-84-7 (Hardcover), 978-981-4364-04-1 (eBook)
www.panstanford.com

5.1 Introduction

5.1.1 *Overview*

The past two decades have witnessed a growing realization of an upcoming energy crisis. Together with an increasing awareness of various environmental issues, people have begun to develop renewable energy sources, such as solar energy, water power, wind power, and biofuels as replacements for traditional fossil fuels and firepower. Solar energy is a particularly promising solution because sunlight is a clean and inexhaustible energy source that can be tapped both on Earth and in outer space. The successful application of solar energy technology might effectively mitigate the degree of global warming caused by unnecessary CO_2 production and environmental pollution. Hence, a staggering amount of research has been conducted in the development of photovoltaic (PV) devices [1–5], systems that convert sunlight directly into electrical power. Among the various kinds of solar cells, organic photovoltaic (OPV) devices based on conjugated-polymer materials are promising systems because they possess many advantageous properties [6, 7], including light weight, mechanical flexibility, low-temperature and low-cost fabrication, and an amazingly short energy payback time, relative to PV cells based on silicon or other inorganic materials [8–12]. Unfortunately, the power conversion efficiencies (PCEs) of OPV devices remain too low for their use in commercial applications. Meanwhile, inferior operating stability of OPVs is another problem that must be solved to ensure their widespread use [13, 14]. Therefore, considerable effort is being exerted toward the goal of commercializing OPVs in the near future [15–20].

This chapter focuses on the development of organic solar cells, especially those based on conjugated polymeric materials. We begin with a discussion of the basic fundamental principles and the electrical characterization of OPVs. Next, we systemically review notable recent developments in the improvement of OPV device performance. In particular, we discuss approaches toward obtaining optimized morphologies of photoactive layers through annealing treatment, and then review developments in electrode interface engineering and the exploitation of optical effects in solar devices.

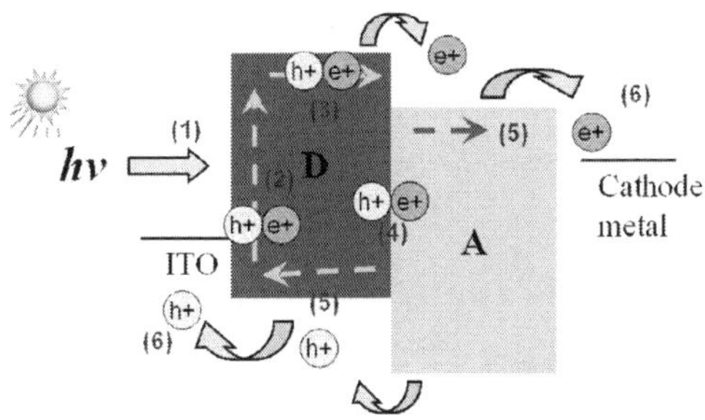

Figure 5.1 Schematic representation of the working principle of an OPV device: (1) light absorption, (2) exciton generation, (3) exciton diffusion, (4) exciton dissociation, (5) charge transport, and (6) charge collection [21]. See also Color Insert.

Significant methods for improving the device operating stability are also introduced. Finally, we highlight the development of novel device structures, such as inverted structures, tandem cells, and stacked cells.

5.1.2 *Basic Principle*

The working principle through which OPVs directly absorb sunlight and then transfer it into electricity involves six steps (Fig. 5.1) [21].

5.1.2.1 Light absorption and collection

This step requires matching of the energy gap of the photoactive materials with the energy of the incident photons. For an effective solar cell, the energy gap of the photoactive layer must be equal to or greater than the energy of the photons from the sunlight. For organic materials, the energy gap corresponds to the difference between the energies of the lowest unoccupied molecular orbital (LUMO) and the highest occupied molecular orbital (HOMO):

$$E_g = \text{HOMO} - \text{LUMO}. \tag{5.1}$$

5.1.2.2 Exciton generation

After light absorption, excitons—pairs of electrons and holes bound through coulombic attraction—are generated in the photoactive layer. Each exciton exists as the combination of an excited and neutral state; the binding energy of an exciton, defined as the energy required to separate the electron/hole pair, typically ranges from 0.1 to 0.5 eV [22, 23].

5.1.2.3 Exciton diffusion

After their generation, the excitons diffuse to the interface of the acceptor and the donor materials within its lifetime. The average distance that an exciton travels is defined as the exciton diffusion length (L_D); for organic materials, the value is ca. 5–70 nm [24].

5.1.2.4 Exciton dissociation

Because of the high binding energies of organic excitons, a lower ionization potential (IP) of the donor material and a higher electron affinity (EA) are necessary for the exciton dissociation process. Under such conditions, the exciton can be separated into free carriers—namely electrons and holes.

5.1.2.5 Charge transport

After the dissociation process, the free electron and hole carriers are transported inside the donor and acceptor materials, respectively, to the electrodes. The transportation of free carriers in organic materials follows the so-called hopping mechanism. Ideally, the free carriers are transported to and then collect by the electrodes. Sometimes, these free carriers might recombine or be trapped by defects; these phenomena contribute to carrier loss, thereby resulting in lower PCEs [25].

5.1.2.6 Charge collection

After transportation of the free carriers to their respective electrodes, the charges are effectively collected and transformed into electricity by the outer electrical circuit. An optimized interface, known as the Ohmic contact, plays an essential role in obtaining superior device performance.

5.1.3 *Device Characterization*

In this sub-section, we briefly describe the parameters used to characterize the performance of PV devices [21]. Figure 5.2a displays a typical current density–voltage (J–V) curve for a solar cell under illumination with sunlight (AM 1.5G, 100 mW/cm^2) (Fig. 5.2b). From this curve, the short-circuit current (J_{sc}), the open-circuit voltage (V_{oc}), and the calculated fill factor (FF) are commonly extracted to characterize the device's performance. The FF of a device is defined as the ratio between the maximum power generated by the device to an external circuit and the potential power ($V_{oc} \times J_{sc}$):

$$FF = \frac{A_1}{A_2} = \frac{P_{max}}{V_{oc} \times J_{sc}} = \frac{V_m \times J_m}{V_{oc} \times J_{sc}} \tag{5.2}$$

where V_m and J_m are the voltage and current density at the maximum power output, respectively. From Fig. 5.2a, the FF can be determined readily from the ratio of the areas of the darkly shaded (A_2) and lightly shaded (A_1) regions. The PCE of a device is defined as the ratio of the maximum electrical power generated (P_{max}) to the incident optical power (P_{in}). The relationship between J_{sc}, V_{oc}, FF, P_{in}, and PCE can be expressed as follows:

$$\text{PCE} = \frac{P_{max}}{P_{in}} = \frac{V_{oc} \times J_{sc} \times FF}{P_{in}} \tag{5.3}$$

In addition to using these parameters, the spectral response of an OPV is another common way to characterize device performance. In

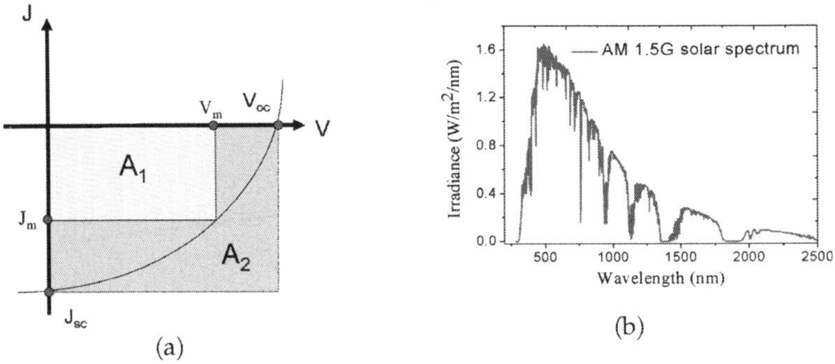

Figure 5.2 (a) J–V characteristics of PV devices operated under illumination. (b) Solar spectrum (AM 1.5G).

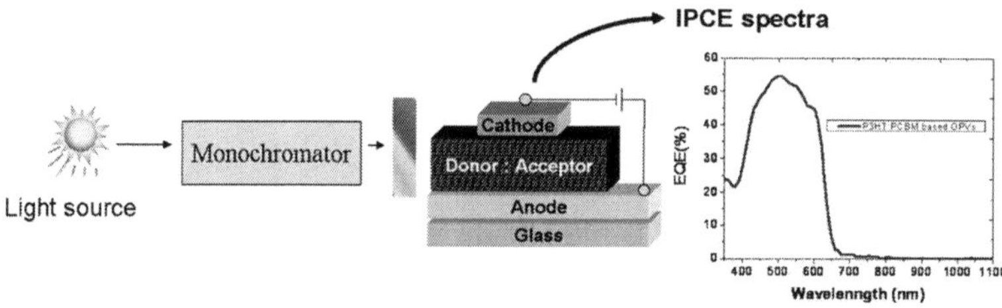

Figure 5.3 Schematic representation of the experimental setup for IPCE measurement.

this approach, as presented in Fig. 5.3, the device is illuminated by a monochromatic light source, generally consisting of a broadband illuminator and a monochromator. The photocurrent is recorded directly as a function of wavelength (λ) and then compared with the light intensity or photo flux. The external quantum efficiency (EQE), also known as the incident photon conversion efficiency (IPCE), is given by the number of electrons (n_e) generated per incident photon (n_{ph}):

$$\text{EQE} = \frac{n_e}{n_{ph}} = 1240 \times \frac{J_{sc}}{\lambda \times I_\lambda} \tag{5.4}$$

where J_{sc} is the photocurrent density (mA/cm^2) under short-circuit conditions, λ is the wavelength (nm), and I_λ is the incident light intensity (mW/cm^2).

5.2 A Brief History of OPVs

In general, small molecule-based and polymer-based solar cells are distinguished by the types of photoactive materials and fabrication methods used. Small molecule-based OPVs are commonly fabricated through thermal evaporation inside a vacuum; their structures feature successive layers of electron- and hole-transporting materials. In 1986, a small molecule-based organic solar cell exhibiting a PCE of ca. 1% was demonstrated by Tang, who employed copper phthalocyanine (CuPc) as the hole-transporting material and perylene derivatives as electron-transporting materials to create a bilayer heterojunction structure (Fig. 5.4a) [1]. In contrast, polymer-

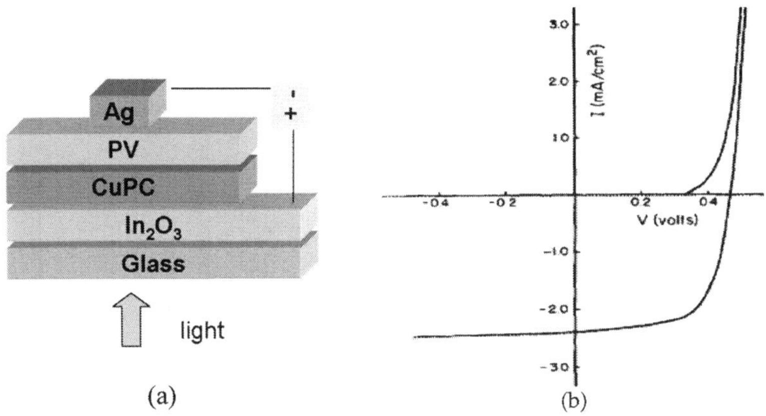

Figure 5.4 (a) Device structure and (b) J–V characteristics of the first PN junction OPV realized by Tang in 1986 [1] (Reprinted with permission from Tang, C. W. (1986). *Appl. Phys. Lett.*, **48**, 183–185; Copyright 1986, American Institute of Physics).

based OPVs are generally prepared using solution processing techniques such as ink-jet printing, screen-printing, spin-coating, dip-coating, and doctor-blade coating [12, 26–28]. Figure 5.5 displays the chemical structures of conducting conjugated polymers that are commonly used as photoactive layers in polymer-based OPVs. Derivatives of polythiophene and polyfluorene commonly serve as *p*-type (electron-donor) materials [29]. All of these materials are solution processable because they are synthesized to present long side-chains. Typical devices consist of blends of a conjugated polymer with soluble small molecules or of blends of two different conjugated polymers. Such blends feature interfaces throughout the photoactive layer, known as the bulk heterojunction (BHJ).

In the initial development of polymer solar cells, devices were constructed based on a bilayer structure, as depicted in Fig. 5.6a. The limited primary area of the heterojunction between the *n*-type (acceptor) and *p*-type (donor) materials results, however, in lower probabilities of photoinduced charge transfer. To overcome this problem, a device structure incorporating mixtures of conjugated polymers and fullerenes as the photoactive layer was reported by G. Yu *et al.* in 1995. As indicated in Fig. 5.6b, such a structure

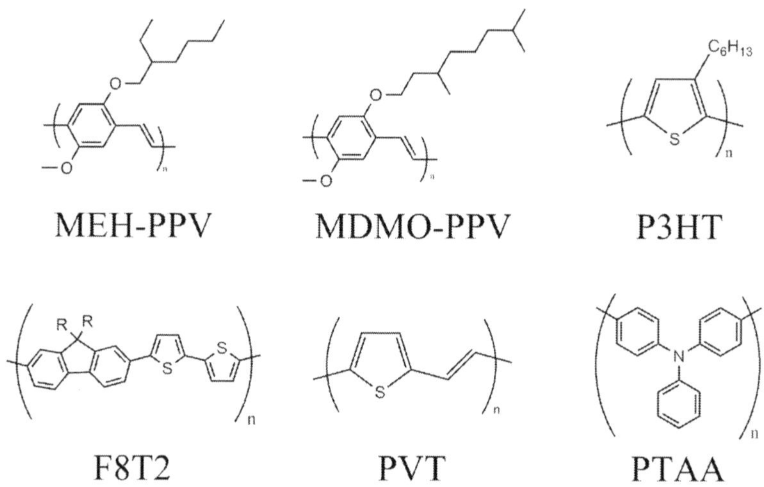

Figure 5.5 Examples of conjugated polymers used in organic solar cells.

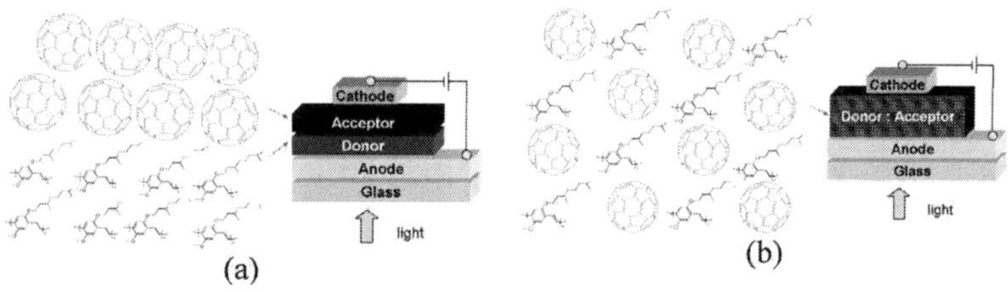

Figure 5.6 Schematic representations of (a) bilayer and (b) BHJ structures in PV devices [30].

possesses a large electron donor/acceptor interface [30]. The use of highly soluble C_{60} derivatives, such as [6,6]-phenyl-C_{61}-butyric acid methyl ester (PCBM), enabled the incorporation of large amounts of these electron donors into the conjugated polymers (Fig. 5.7) [30]. This approach provided a carrier collection (η_c) of 29% and a PCE (η_e) of 2.9% [30].

Furthermore, in addition to the choice of electron acceptor materials, the optical and electrical characteristics of the donor materials also significantly influence the efficiencies of OPVs. Considerable progress has been made recently toward developing superior donor materials providing highly efficient absorption and high carrier mobility. For instance, in 2002, Schilinsky prepared BHJ

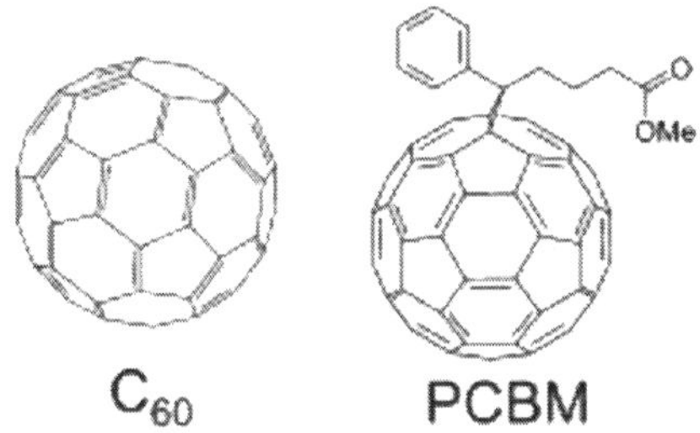

Figure 5.7 Chemical structures of buckminsterfullerene (C_{60}) and its soluble derivative PCBM [32].

devices from mixtures of P3HT and PCBM and achieved comparable levels of performance [31]. The device structure consisted of a conducting polymer layer of PEDOT:PSS, which was spin-coated on indium tin oxide (ITO)–coated glass. Subsequently, a layer of P3HT and PCBM was deposited on the PEDOT:PSS layer. The cathode metals on the top of the device comprised calcium capped by silver. For OPVs based on P3HT:PCBM blends, the photocurrents are higher than those of PPV-based devices (8.7 mA/cm2) because of the higher absorption and better energy-match between the absorption of P3HT and the solar spectrum, even though the open-circuit voltage is rather low (0.58 V). P3HT-based OPVs can exhibit a PCE of 2.8%, an FF of 55%, and a peak EQE of 70%, outstanding performance relative to that of PPV-PCBM devices [31]. In summary, the most highly efficient polymer-based PV cells are fabricated based on the BHJ structure, generally consisting of a light-absorbing polymer as the electron donor and a soluble C_{60} derivative as the electron acceptor.

The morphology of the photoactive layer dramatically influences the performance of organic solar cells based on polymer materials, including their light absorption ability, exciton generation, exciton dissociation, charge transport, and charge collection processes. Some remarkable approaches, such as pre-thermal annealing and so-called "solvent annealing," have been invented to enhance the efficiency of OPVs [19, 33, 34]. For example, in 2005, Li *et al.* improved the device performance of P3HT/PCBM solar cells through solvent annealing [32]. After spin-coating, the wet active layer was covered with a glass Petri dish to create a solvent-saturated environment.

Because of the slow rate of evaporation of solvent molecules under such conditions, the polymer layer slowly solidified. As a result, the photoactive layer featured a higher degree of ordering, leading to a higher photon absorption ability, superior carrier transportation, and better carrier collection; as a result, the PCE improved to as high as 4.4% [33]. In recent years, polymer/fullerene BHJ solar cells have been prepared using a variety of annealing processes [34–36], which have been summarized in recent review articles [37, 38]. In the following section, we discuss more recent progress, focusing on a series of effective approaches for enhancing the efficiency and device stability of polymeric OPVs.

5.3 Current Approaches Toward Enhancing the Performance of OPVS

As mentioned above, high-efficient polymer solar cells are typically based on a BHJ architecture comprising a blend of P3HT and PCBM. Although the morphology plays an essential role in determining the device performance of polymer/fullerene solar cells, several questions remain—for example: how can we enhance the absorption ability using optical approaches? In addition, there remains a need to enhance device stability. Therefore, in this section we summarize recent developments in the search for higher efficiency and superior device stability.

5.3.1 *Electrode Modification and Engineering*

The performance of OPVs is dramatically influenced by the electrical properties of the electrodes, such as the conductivity and the quality of the interfacial contacts of the anode and cathode. Modification of electrodes is commonly employed to improve the device performance of OPVs. For example, Ko *et al.* doped mannitol, a polyol, into the PEDOT:PSS layer; the resulting increased conductivity of the PEDOT:PSS layer improved the device efficiency (Fig. 5.8; Table 5.1) [39]. The enhancement in device performance could be attributed to the reduced resistance of the PEDOT:PSS layer after blending with mannitol, thereby effectively minimizing the

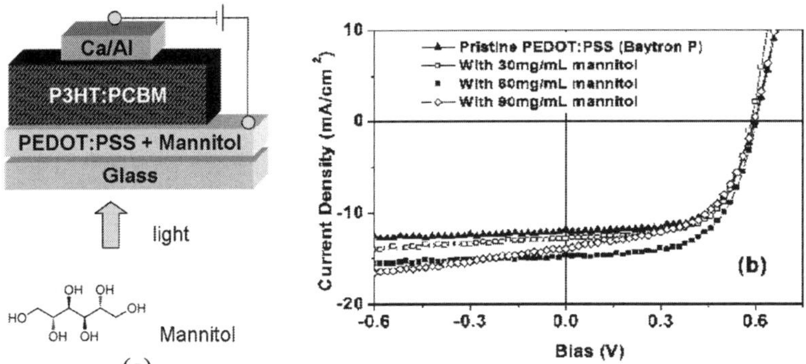

(a) (b)

Figure 5.8 (a) Device structure employed by Ko *et al.* (b) J–V curves of polymer PVs featuring different concentrations of mannitol-doped anodes: without doping (▲) and with doping at 30 (□), 60 (•), and 90 mg/mL (○); under of AM 1.5G illumination (100 mW/cm^2) [39]. (Reprinted with permission from Ko, C. J., Y. K. Lin, F. C. Chen, Chu, C. W. (2007). Modified buffer layers for polymer photovoltaic devices, *Appl. Phys. Lett.*, **90**, 063509; Copyright 2007, American Institute of Physics.)

Table 5.1 Performances of OPVs featuring various mannitol concentrations in their PEDOT:PSS layers; under AM 1.5G illumination (100 mW/cm^2) [39]

Anode	V_{oc} (V)	J_{sc} (mA/cm^2)	FF (%)	PCE (%)	R_s^a (Ω cm^2)	Resistivity[b] (Ω cm)
Pristine PEDOT:PSS	0.6	12	63.4	4.5	1.9	20.29
30 mg/mL mannitol in PEDOT:PSS	0.59	13.1	63.2	4.9	1.42	3.72
60 mg/mL mannitol in PEDOT:PSS	0.59	14.7	60.1	5.2	1.29	0.53
90 mg/mL mannitol in PEDOT:PSS	0.59	13.8	55.2	4.5	1.01	0.16

[a]Calculated from the I–V curves recorded in the dark.
[b]Measured using a four-point probe.
Reprinted with permission from Ko, C. J., Y. K. Lin, F. C. Chen, Chu, C. W. (2007). *Appl. Phys. Lett.*, **90**, 063509 Copyright 2007, American Institute of Physics.

device's series resistance. Following this approach, the best device performance, featuring a high PCE of 5.2%, occurred when the mannitol concentration in the PEDOT:PSS solution was 60 mg/mL.

This result suggests that the conductivity of PEDOT:PSS has a dramatic influence on the device performance, in terms of the values of J_{sc}, FF, and PCE. Therefore, the bulk resistance of the photoactive layer is not the only feature to take into consideration: the use of a PEDOT:PSS buffer layer exhibiting a suitable high conductivity is also essential for the development of highly efficient polymer-based solar cells.

The device performance can be also improved by altering the cathode structure to obtain a lower barrier height between the photoactive layer and the cathode. In 2008, Yip *et al.* improved the device performance by inserting a layer of ZnO, modified with self-assembled monolayers (SAMs), between the photoactive layer and the metal cathode [40]; Fig. 5.9a displays the device architecture of this polymer solar cell, comprising a ZnO/SAM/metal cathode, and the chemical structures of the molecules used to form the SAMs (lauric acid (LA), mercaptoundecanoic acid (MUA), and perfluorotetradecanoic acid (PFTDA)). As illustrated in Fig. 5.9b, incorporating both LA and MUA led to interfacial dipoles pointing away from the metal surface, thereby resulting in a reduction of the metal work function and, therefore, an Ohmic contact (Fig. 5.9b-iii). Notably, modifications based on MUA or LA as an interfacial modifier resulted in higher device performance. On the other hand, devices modified with PFTDA exhibited low values of V_{oc} and poor OPV characteristics as a result of the formation of an unfavorable interfacial dipole (Fig. 5.9b-ii). In short, Yip *et al.* found that selecting appropriate SAMs can improve the quality of the interface between the photoactive layer and the metal cathode; as a result, they obtained high efficiencies (4.3–4.6%). Furthermore, they demonstrated that high-work function metals (Ag, Al, Au) could also be used as satisfactory cathodes in polymer solar cells.

More recently, we developed a conceptually new method for interface modification toward preparing self-organized bilayer structures in polymer solar cells [41]. In this approach, we introduced a nanoscale functional interlayer, containing poly(ethylene glycol) (PEG), through spontaneous vertical phase separation to enhance the efficiency and increase the stability of polymer solar cells. Figure 5.10a illustrates the procedures used for the fabrication of polymer solar cells featuring a self-organized bilayer

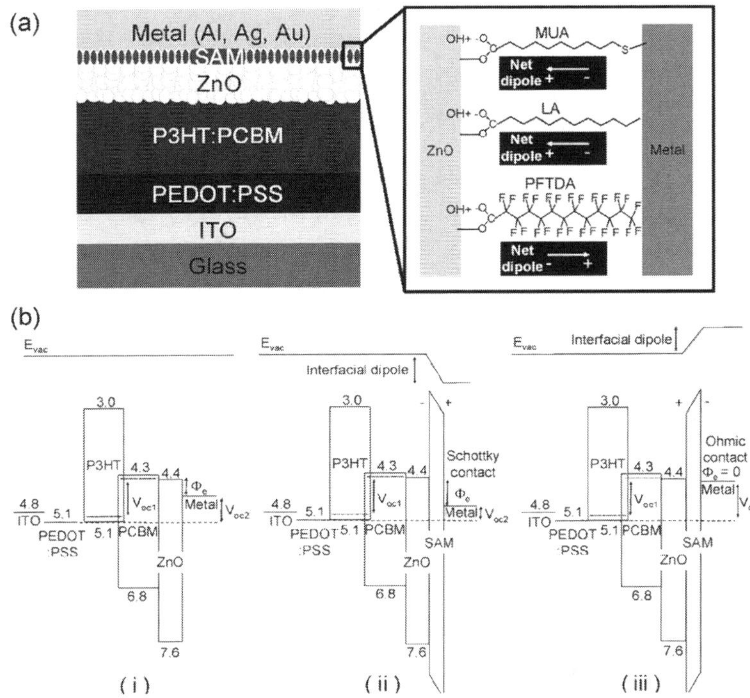

Figure 5.9 (a) Device architecture of polymer solar cells featuring ZnO/SAM/metal cathodes, prepared by Yip *et al.*; chemical structures and dipoles of the SAMs. (b) Schematic representation of the energy levels of devices incorporating (i) a ZnO/metal cathode, (ii) a ZnO/SAM/metal cathode with interfacial dipoles directed toward the metal surface, and (iii) a ZnO/SAM/metal cathode with interfacial dipoles directed away from the metal surface [40] (Reprinted with permission from Yip, H. L., S. K. Hau, N. S. Baek, Jen, A. K. Y. (2008). *Appl. Phys. Lett.*, **92**, 193313; Copyright 2008, American Institute of Physics).

(P3HT:PCBM/PEG) structure. Scanning electron microscopy (SEM) images revealed morphological evidence supporting the formation of vertically-phase-separated PEG molecules (Fig. 5.10b). A film containing no PEG exhibited a smooth surface morphology. After blending 5% or 10% PEG into the film, "dot-like" phases were present on the surfaces of the films, with features having an average diameter of ca. 30 nm. As indicated in Fig. 5.10c, the addition of PEG (5 wt%) into the active layer improved the device performance considerably in terms of its values of V_{oc}, J_{sc}, and FF, thereby enhancing the PCE from 2.21 to 3.97%. We attribute this improved performance to interactions between PEG and Al that reduced the injection barrier height and contact resistance between the polymer layer and the cathode metal and, consequently, optimized electron collection. In addition, OPVs fabricated using this approach exhibited superior device operating stability under illumination (AM1.5G, 100 mW/cm^2), relative to that of corresponding devices incorporating Ca/Al or LiF/Al structures, which are commonly used as cathodes of OPVs (Fig. 5.10d). In summary, this approach provides an effective interface structure that results in not only highly efficient but also highly stable devices. Note that only a single step is required to fabricate both the active and cathode buffer layers, thereby simplifying the device fabrication process.

5.4 Optical Effects

One of the key issues toward achieving high device efficiency is sufficient photo-absorption in the photoactive layer, i.e., efficient harvesting of sunlight. The common approach is to use a thicker active layer; the use of too thick films, however, inevitably results in increased device resistance, due to the low carrier motilities of organic materials. An alternative approach toward enhanced light absorption without using thick layers is the exploitation of optical effects. In this section, we summarize recent remarkable approaches for increasing the efficiency of light absorption.

5.4.1 *Surface Plasmonic Effects*

One of strategies for increasing the light absorption of OPVs is to incorporate metal nanoparticles (NPs), which can produce localized surface plasmon resonance (LSPR) upon illumination, into the devices. The metallic nanostructure can induce a large electromagnetic field enhancement around the metal surface, and, therefore, effectively increase the optical electrical field inside the photoactive layer, thereby enhancing optical absorption of the photoactive layer and providing superior photoconversion near the region of the resonance wavelengths. More recently, several groups

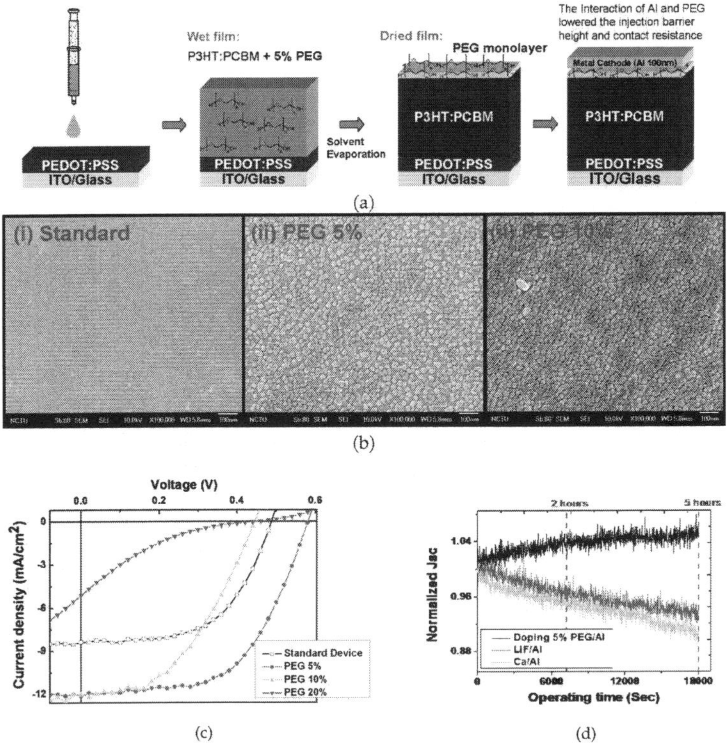

Figure 5.10 (a) Schematic representation of the fabrication of solar cells featuring a self-organized bilayer (P3HT:PCBM/PEG) structure. (b) SEM images of the surfaces of active layers prepared at PEG concentrations of (i) 0, (ii) 5, and (iii) 10%. (c) $J-V$ plots of devices prepared with and without PEG, measured under simulated AM1.5G illumination (100 mW/cm^2). (d) Photocurrent stability measurements of devices incorporating various cathode structures, under AM1.5G illumination (100 mW/cm^2) for 5 h. The initial values of J_{sc} were normalized for ready comparison. The active layers of the devices incorporating LiF/Al and Ca/Al as cathode materials did not contain any PEG [41] (Chen, F. C., Chien, S. C. (2009). *J. Mater. Chem.*, **19**, 6865–6869; Reproduced by permission of the Royal Society of Chemistry). See also Color Insert.

have reported increased photocurrents after exploiting this optical phenomenon using different metallic NPs. For example, Morfa *et al.* incorporated Ag NPs, prepared through thermal evaporation, onto ITO-coated glass substrates [42], to form devices having the

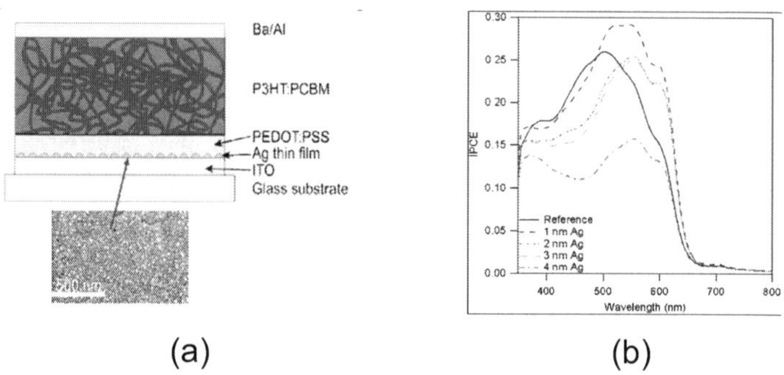

Figure 5.11 (a) Structure of a device incorporating a thin Ag film deposited onto the ITO-coated glass substrate. Inset: Field-emission SEM micrograph of a representative 2-nm-thick Ag layer on ITO. (b) IPCE spectra of devices containing 1-, 2-, 3-, and 4-nm-thick Ag films [42] (Reprinted with permission from Morfa, A. J., K. L. Rowlen, T. H. Reilly, M. J. Romero, van de Lagemaat, J. (2008). *Appl. Phys. Lett.*, **92**, 013504, Copyright 2008, American Institute of Physics).

structure ITO-coated glass/Ag NPs/PEDOT:PSS/P3HT:PCBM/Ba/Al (Fig. 5.11a). Table 5.2 reveals that the resulting solar energy conversion efficiency for a BHJ PV device increased from 1.3 (±0.2) to 2.2 (±0.1)% after incorporating a 1- or 2-nm-thick layer of plasmon-active Ag. IPCE spectra revealed (Fig. 5.11b) a strong increase in EQE at longer wavelengths for all Ag-containing devices, except for that having a nominal Ag film thickness of 4 nm.

Table 5.2 Average device parameters as a function of the nominal thickness of the Ag film; errors reported as one standard deviation [42]

Ag height (nm)	J_{sc} (mA/cm^2)	η (%)	V_{oc} (mA)
0 (Ref)	4.6 ± 0.4	1.3 ± 0.2	566 ± 5.6
1	6.9 ± 0.2	2.2 ± 0.1	590 ± 5.8
2	7.3 ± 0.4	2.1 ± 0.1	581 ± 8.8
3	6.5 ± 0.1	1.8 ± 0.2	564 ± 30.9
4	2.6 ± 0.4	0.9 ± 0.1	599 ± 6.2

Reprinted with permission from Morfa, A. J., K. L. Rowlen, T. H. Reilly, Romero, M. J., van de Lagemaat, J. (2008). *Appl. Phys. Lett.*, **92**, 013504. Copyright 2008, American Institute of Physics.

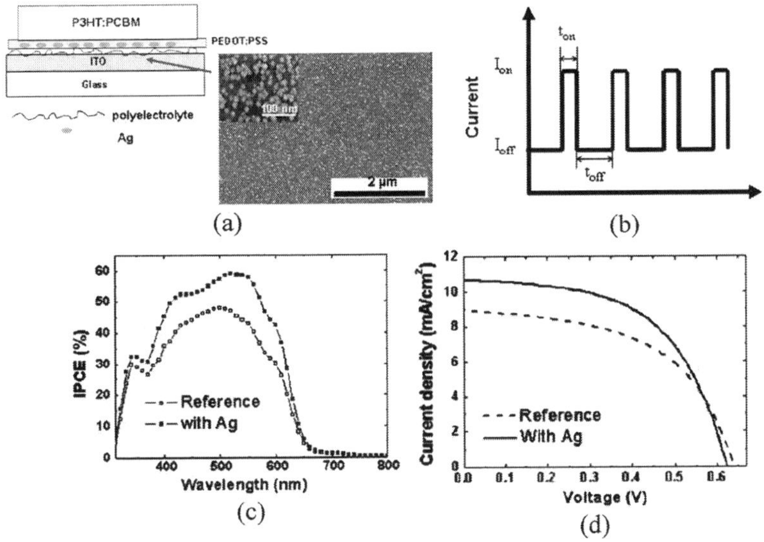

Figure 5.12 (a) Structure of a polymer solar cell featuring Ag NPs. (b) Conceptual approach of using PC current electrodeposition to prepare Ag NPs. (c) IPCE spectra and (d) J–V curves of polymer solar cells with and without Ag NPs [43] (Reprinted with permission from Kim, S. S., S. I. Na, J. Jo, D. Y. Kim, Nah, Y. C. (2008). *Appl. Phys. Lett.*, **93**, 073307, Copyright 2008, American Institute of Physics).

Another approach, reported by Kim *et al.*, for incorporating surface-plasmonic Ag NPs into the devices involves the use of pulse-current (PC) electrodeposition [43]. With this method, the size, density, and morphology of the NP films can be controlled readily (Fig. 5.12a). Under the electrodeposition conditions described in Fig. 5.12b, Kim *et al.* fabricated relatively uniform Ag NPs having an average particle size of ca. 13 nm. After incorporating these plasmonic Ag NPs, IPCE spectra revealed an enhancement over the entire excitation spectral range longer than ca. 400 nm (Fig. 5.12c), thereby increasing the PCE from 3.05 to 3.69% (Fig. 5.12d).

In addition to using Ag NPs, Au nanostructures are also candidates for producing LSPR phenomena to achieve highly efficient polymer solar cells. In 2009, Lee *et al.* reported the enhanced performance of P3HT:PCBM BHJ solar cells prepared with wet-deposited interfacial gold nanostructures on ITO-coated substrates

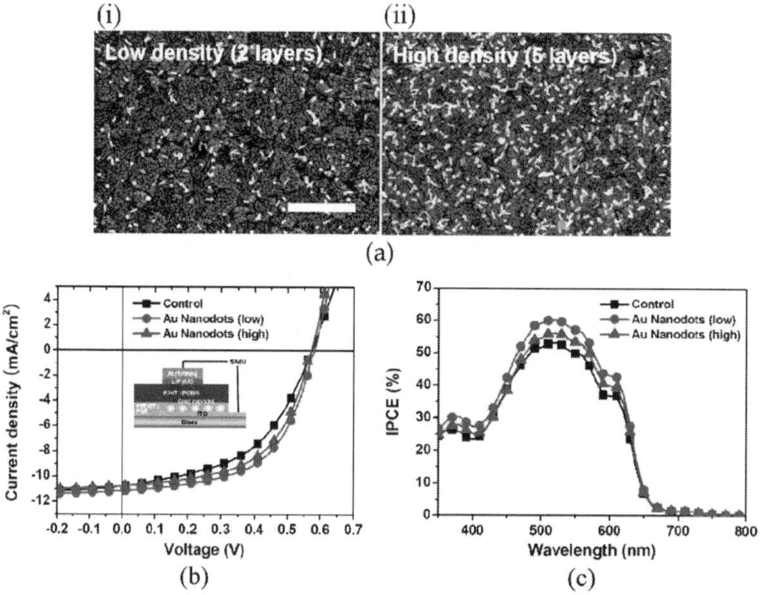

Figure 5.13 (a) SEM images of Au nanrods in (i) low-density (two layers) and (ii) high-density (five layers) configurations. Scale bar: 500 nm. (b) *J–V* curves of polymer solar cells with and without Au nanodots, recorded under AM1.5 simulated light (100 mW/cm^2). (c) IPCE spectra of these devices. The device structure with Au nanodots is shown in the inset to (b) [41] (Reprinted from Lee, J. H., J. H. Park, J. S. Kim, D. Y. Lee, Cho, K., High efficiency polymer solar cells with wet deposited plasmonic gold nanodots, *Org. Electron.*, **10**, 416–420, Copyright (2009), with permission from Elsevier).

[44]. In particular, the Au nanorods were prepared through layer-by-layer electrostatic assembly, with the particle density controlled by varying the number of deposition cycles. Figure 5.13a displays the difference in the particle densities of Au nanorods in two and five layers, defined as low- and high-density conditions, respectively. Figure 5.13b reveals that the device containing a low density of Au nanodots provided a higher PCE (3.65%) relative to that of the control device (3.04%). IPCE spectra (Fig. 5.13c) supported the notion that the device enhancement was resulted from the low density of Au NPs.

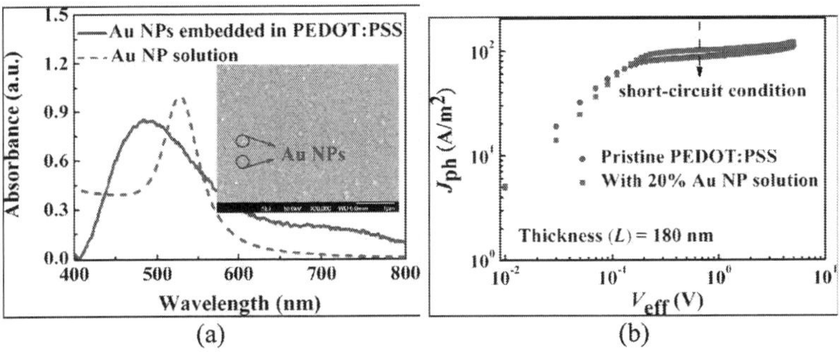

Figure 5.14 (a) Absorption spectra of a Au NP solution and of Au NPs embedded in PEDOT:PSS. Inset: SEM image of the PEDOT:PSS film prepared with 20% Au NP solution blended into the matrix. (b) Plots of photocurrent density ($J_{\rm ph}$) with respect to the effective bias ($V_{\rm eff}$) for a reference device and for a device containing 20% Au NP solution [45] (Reprinted with permission from Chen, F. C., J. L. Wu, C. L. Lee, Y. Hong, C. H. Kuo, Huang, M. H. (2009). *Appl. Phys. Lett.*, **95**, 013305, Copyright 2009, American Institute of Physics).

More recently, we reported a convenient approach for incorporating surface-plasmonic metal NPs into polymer solar cells [45]. This process for fabricating devices simply involved the blending of Au NPs and PEDOT:PSS solutions. A nanocomposite layer was formed after spin-coating such a solution mixture (Fig. 5.14a). The size of the NPs (30–40 nm) was intentionally selected so that their SPR peak matched the absorption wavelength of P3HT. Table 5.2 compares the device performance obtained for this system at various concentrations of Au NPs. We observed a significant increase in the value of $J_{\rm sc}$ after incorporation of the Au NPs. The optimal device featured a value of $J_{\rm sc}$ of 10.18 mA/cm^2, leading to a superior PCE of 4.19%, relative to that of the reference cell. On the other hand, Table 5.2 reveals that both the calculated device series resistance ($R_{\rm s}$) and the measured resistivity of the PEDOT:PSS buffer layer increased after the addition of the Au NPs, suggesting that the enhancement in $J_{\rm sc}$ was not due to the improved device resistance. To further understand the mechanism of operation, we also calculated the maximum exciton generation rates ($G_{\rm max}$) and exciton dissociation probabilities, which are closely related to the degree of light harvesting and the recombination

Table 5.2 Performance of OPVs incorporating PEDOT:PSS doped with various concentrations of Au NPs [45]

	V_{oc} (V)	J_{sc} (mA/cm)	FF (%)	R_s^a (Ω-cm²)	Resistivity[b] (Ω-cm)	PCE (%)
Pristine PEDOT:PSS	0.59	8.95	65.9	1.73	32	3.48
10% Au NP solution	0.59	9.5	69.2	1.86	37.1	3.89
20% Au NP solution	0.59	10.18	69.8	2.08	43.5	4.19
30% Au NP solution	0.59	9.46	69.3	2.35	47.5	3.87

[a] Device series resistance of the OPV was obtained from the inverse slope of the dark $J-V$ curve at a voltage of 1.5 V.
[b] Resistivity of the buffer layer determined using the four-point probe method.
Reprinted with permission from Chen, F. C., J. L. Wu, C. L. Lee, Y. Hong, C. H. Kuo, Huang, M. H. (2009). *Appl. Phys. Lett.*, **95**, 013305. Copyright 2009, American Institute of Physics.

rate, respectively [46, 47]. After incorporation of the metal NPs, the values of G_{max} and the dissociation probabilities were drastically enhanced—from 3.82×10^{27} to 4.13×10^{27} m^{-3}s^{-1} and from 78.7 to 83.9%, respectively—thereby resulting in a higher photocurrent and FF. This approach is a simple means of exploiting plasmonic-enhanced effects in polymer solar cells. Furthermore, such low temperature processing enables the application of metal NPs on flexible substrates.

5.4.2 *Light Trapping by Periodic Grating Structure*

Another strategy toward effectively harvesting sunlight and enhancing the photocurrents from OPVs is to increase the path length and, thereby, trap photons in the photoactive layer. Using this approach, in 2009, submicron periodic structures were incorporated into polymer solar cells to form an efficient light-trapping scheme [48]. The grating structure of the active layer, as well as the cathode, was fabricated using a damage-free soft lithography approach, in which photoinduced surface-relief gratings (SRGs) on azo-polymer films and poly(dimethylsiloxane) were used as the master and stamp, respectively (Fig. 5.15). Figure 5.15a displays the device fabrication procedure, which is involved formation of the active layer, soft contact printing, detachment of the PDMS mold, and cathode evaporation. Figure 5.15b presents the AFM morphologies of the surfaces of the active layer, the reflective cathode of the

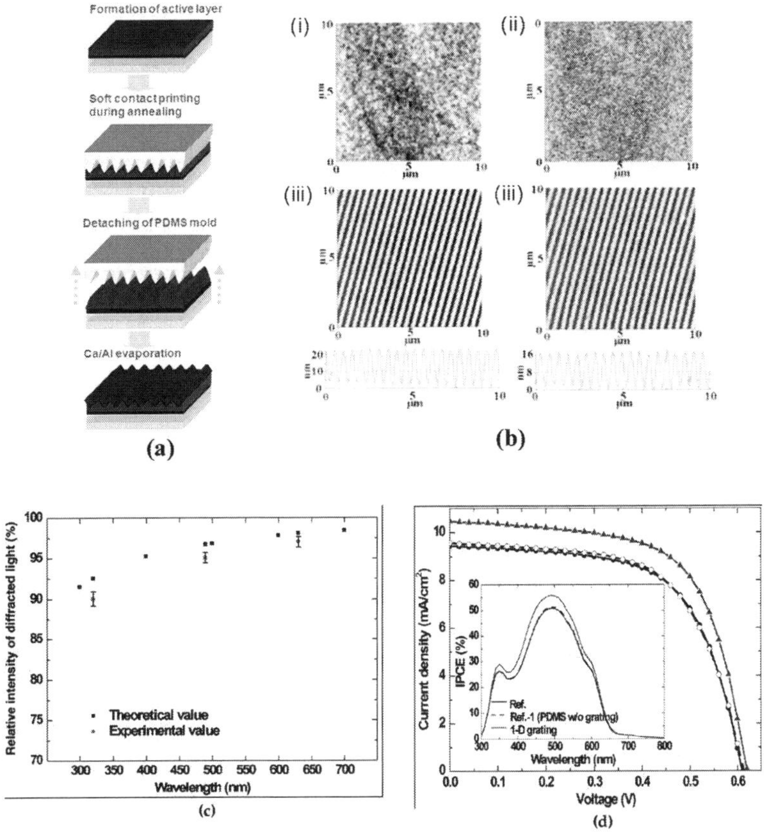

Figure 5.15 (a) Schematic representation of the fabrication of polymer solar cells featuring SRGs. (b) AFM image of (i) the flat active layer of the reference cell, (ii) the reflective cathode of the reference cell, (iii) the active layer patterned via soft contact printing, and (iv) the reflective cathodes evaporated onto the patterned active layer. (c) Diffraction efficiency plotted as a function of wavelength; the intensity of diffracted light was normalized with respect to that of the corresponding cell lacking the grating. (d) J–V characteristics for three kinds of solar cells: a reference cell (closed circles), a second reference cell treated with the PDMS mold but lacking SRGs (open circles), and a cell treated with the PDMS mold and featuring SRGs (closed triangles). Inset: IPCE spectra for the three cells [48] (Reprinted with permission from Na, S. I., Seok-Soon, K., S. S. Kwon, J. Jang, K. Juhwan, T. Lee, Dong-Yu, K. (2007). *Appl. Phys. Lett.*, **95**, 013305, Copyright 2007, American Institute of Physics).

reference cells, and the patterned cells. To confirm the existence of light-bending phenomena, the diffraction efficiency was investigated quantitatively as a function of the wavelength ($\lambda = 325$, 488, 632 nm). Figure 5.15c reveals the experimentally measured diffraction efficiency in the region between 300 and 700 nm, as well as the theoretical results; an increase in the optical path as a result of the diffraction of light was obtained directly over a broad range, especially in the device absorption range. To further confirm the effect of this grating structure on the performance of polymer solar cells, photocurrent density–voltage (J –V) curves were measured and the incident photon-to-current conversion efficiency (IPCE) was calculated Fig. 5.15d. Optimally, the device efficiency was enhanced from 3.56 to 4.11% when using this structure. These observations imply that the periodic structure effectively induced more photogenerated charge carriers through stronger absorption of the photoactive layer, resulted from the increased optical pass length and increased degree of light trapping [45].

5.4.3 *Redistribution of Optical Electric Field in Photoactive Layer*

Figure 5.16a reveals that, according to the results of optical simulation, another critical issue affecting the performance of OPVs is the distribution of the optical electric field in the photoactive layer. For conventional devices, the field distribution is relatively low in the photoabsorption layer, resulting in a lower rate of exciton generation. Moreover, more electron/hole pairs produced near the ITO/PEDOT:PSS electrode will result in low efficiency because the excitons may be quenched by the electrode [49]. To overcome these problems, Kim *et al.* proposed the introduction of an optical spacer between the active layer and the Al electrode, as illustrated in Fig. 5.16b [50]. They used of a solution-based sol–gel process to fabricate a titanium oxide (TiO_x) layer on top of the photoactive layer to function as an optical spacer. After incorporating the optical spacer, the optical-electric field in the photoactive layer was modified, with the maximum signal of the distribution tuned to fall into the photoactive layer (Fig. 5.16a). The IPCE spectra of these devices exhibited substantial enhancements (ca. 40%) over

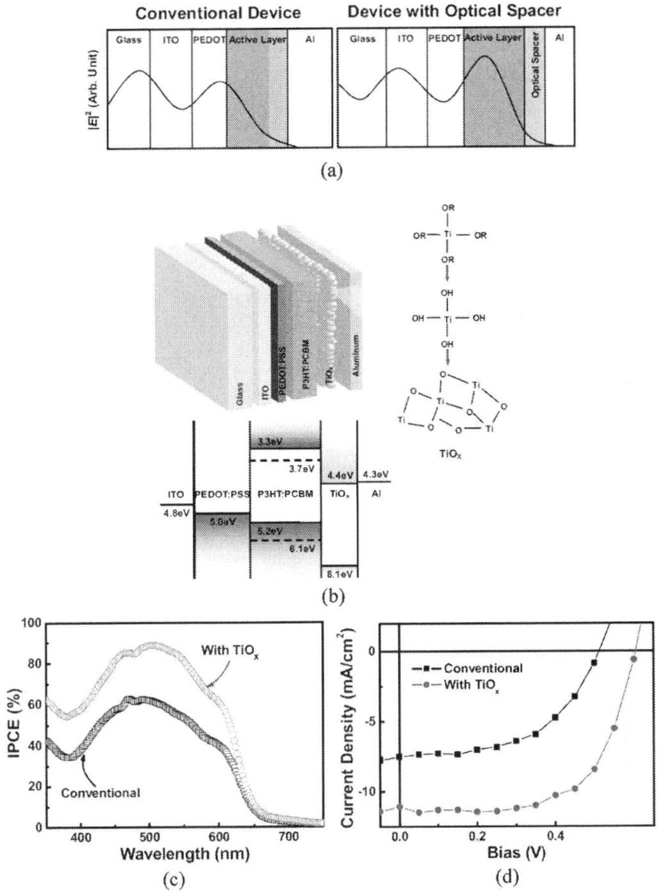

Figure 5.16 (a) Schematic representation of the spatial distribution of the optical electric field strength $|E|^2$ inside devices with (top) and without (down) an optical spacer. (b) Schematic representation of the device structure; a brief flow chart of the steps involved in the preparation of the TiO_x layer; and the energy levels of ITO, PEDOT:PSS, P3HT, PCBM, TiO_x, and Al. (c) Incident monochromatic photon-to-current collection efficiency spectra for devices with and without the TiO_x optical spacer layer. (d) Current density–voltage characteristics of polymer solar cells with (circles) and without (squares) the TiO_x optical spacer under AM1.5 illumination from a calibrated solar simulator (90 mW/cm^2) [50] (Kim, J. Y., S. H. Kim, H. H. Lee, K. Lee, W. L. Ma, X. Gong, Heeger, A. J. (2006). New architecture for high-efficiency polymer PV cells using solution-based titanium oxide as an optical spacer, *Adv. Mater.*, 18, 572, Copyright Wiley-VCH Verlag GmbH & Co. KGaA. Reproduced with permission).

the entire excitation spectral range (Fig. 5.16c). This improvement can be attributed to the increased number of photogenerated charge carriers that resulted from the optimized redistribution of the light intensity. In terms of performance (Fig. 5.16d), the device featuring the TiO_x layer was superior ($J_{sc} = 11.1$ mA/cm^2; $V_{oc} = 0.61$ V; FF = 0.66; PCE = 5%) to the conventional device ($J_{sc} = 7.5$ mA/cm^2; V_{oc} = 0.51 V; FF = 0.54; PCE = 2.3%). Moreover, the increased IPCE was consistent with the trends found for the values of J_{sc} (i.e., a ca. 50% increase for the TiO_x device).

In addition to the earlier demonstration of the effect of optical spacers, Gilot *et al.* performed a complementary theoretical/experimental study in which a layer of optical spacer (ZnO) was inserted between the active layer and the reflective electrode (Fig. 5.17a) [51]. Figure 5.17 presents the calculated optical electrical field distributions at wavelengths of 400, 500, and 600 nm. The thickness of the P3HT:PCBM layer was 40 nm. In the absence of the optical spacer layer, the maximum electric field was located beyond the photoactive layer, indicating that light absorption was not optimized yet. Insertion of a 39-nm-thick ZnO layer shifted the position of the maximum into the photon-absorbing layer. These calculations suggested that a higher photocurrent would be obtained after insertion of an optical spacer. To quantify this effect, the maximum value of J_{sc} was calculated under the assumption that the internal quantum efficiency was up to 100% when the device was illuminated at 100 mW/cm^2 (AM1.5G). Figure 5.17c reveals that there was an almost two-fold increase in IPCE after adding a 40-nm-thick ZnO layer to a device containing a 40-nm-thick active layer. Unexpectedly, cells incorporating a larger film thickness (70–130 nm) provided contrasting results. Gilot *et al.* speculated that the active layer was already optimized and that the maximum of the electric field intensity shifted away from the active layer after incorporating an optical spacer. To further understand this optical effect, they fabricated devices featuring photoactive layers of various thicknesses (27–220 nm), in either the presence or absence of a 39-nm-thick ZnO spacer layer. The experimental results (Fig. 5.17d) agreed well with the calculated data. The lower photocurrent was due to the fact that the internal quantum efficiency of real devices is less than 100%. Therefore, the authors concluded that

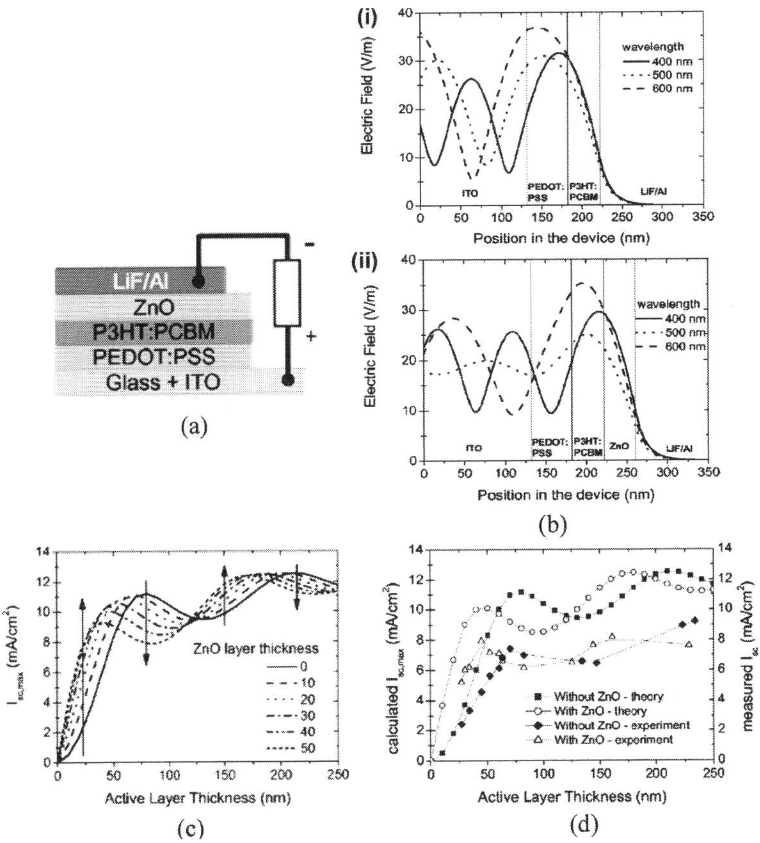

Figure 5.17 (a) Structure of a device featuring a ZnO optical spacer. (b) Simulated results for the optical electric field distributions of three different wavelengths of light for devices incorporating a 40-nm-thick active layer (i) without and (ii) with a 39-nm-thick ZnO layer. (c) Simulated active layer thickness/short circuit current curves for devices with various ZnO layer thicknesses [assuming 100% IQE and AM1.5 illumination (100 mA/cm^2)] (d) Theoretical and experimental plots of short circuit current with respect to the active layer thickness, with or without a 39-nm-thick ZnO optical spacer [51] (Reprinted with permission from Gilot, J., I. Barbu, M. M. Wienk, Janssen, R. A. J. (2007). *Appl. Phys. Lett.*, **91**, 113520, Copyright 2007, American Institute of Physics).

the absorption efficiency could be enhanced by incorporating an optical spacer when the thickness of the active layer was less than ca. 60 nm. For thicker layers, however, insertion of a ZnO layer might not enhance the photocurrent. The effect of such an optical spacer depends strongly on the thickness of the photoactive layer.

5.4.4 *Spectral Coverage Enhancement*

Compared with the sun's broad spectrum, organic materials usually have limited spectral coverage. Many approaches have been proposed recently to overcome this problem. For example, materials having a lower band gap can be used as the photoactive layer; alternatively, multijunction devices, which contain semiconductors with different band gaps for sequentially extracting sun power from their respective absorption range, can be fabricated. Here, we first review one approach to extend the absorption spectral coverage by doping with molecules exhibiting near-infrared absorption. An alternative approach, through construction of multijunction devices, will be presented in the next section.

In 2008, Peet *et al.* incorporated a soluble small molecule donor (SMD) into polymer solar cells to extend the absorption spectrum of the photoactive layer to 800 nm [52]. Figure 5.18a presents the chemical structures of the components in the active layer and their energy levels. To avoid charge trapping on the dyes, the energy levels of the low-bandgap chromophore should lie between those of the polymer and the fullerene. Figure 5.18b displays the individual optical absorption spectra of thin films of P3HT, C_{71}-PCBM, and SMD1; Fig. 5.18c provides the absorption spectra of P3HT:PCBM films containing SMD1 (0, 2, and 6 mg/mL). The addition of SMD1 into the P3HT:PCBM films increased the absorption, especially in the region 650–800 nm. As presented in Fig. 5.18d, the device efficiency was enhanced from 2.93 to 3.21% after blending SMD1 at 2 mg/mL into the active layer, because of the increased number of photogenerated carriers. The corresponding IPCE spectra (Fig. 5.18e) reveal that dye absorption near 700 nm contributed to the device photocurrent for the samples containing SMD1. Most importantly, this approach demonstrates that the incorporation of soluble small molecule sensitizers exhibiting

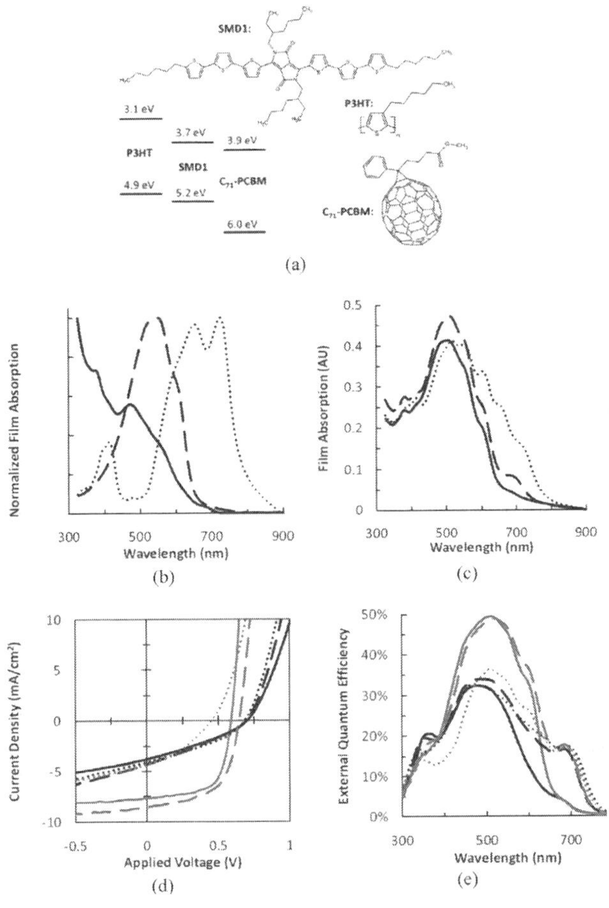

Figure 5.18 (a) Structures of the three solar cell components and their approximate energy levels. (b) Normalized optical absorption spectra for thin films of C_{71}-PCBM (solid), P3HT (dashed), and SMD1 (dotted). (c) Optical absorption spectra of BHJ blend films comprising P3HT:C_{71}-PCBM (10:10; solid), P3HT:C_{71}-PCBM:SMD1 (10:10:2; dashed), and P3HT:C_{71}-PCBM:SMD1 (dotted) after annealing. (d) $J-V$ curves and (e) EQE spectra of BHJ solar cells incorporating P3HT:C_{71}-PCBM (10:10; solid), P3HT:C_{71}-PCBM SMD1 (10:10:2; dashed), and P3HT: C_{71}-PCBM:SMD1 (10:10:6; dotted), recorded before (black) and after (gray) thermal annealing [52] (Reprinted with permission from Peet, J., A. B. Tamayo, X. D. Dang, J. H. Seo, Nguyena, T. Q. (2008). *Appl. Phys. Lett.*, **93**, 163306, Copyright 2008, American Institute of Physics).

near-infrared absorption into devices is a potential path toward efficient light harvesting at longer wavelengths.

5.4.5 *Novel Device Structures*

Many novel device structures, including inverted structures, stacked structures, and tandem cells, have been prepared in the search for superior device characteristics with simpler, lower cost fabrication. These new device designs can provide higher efficiencies, more stable operation, and longer storage stability under ambient conditions; in addition, their fabrication processes are more compatible with flexible substrates. In this section, we discuss the properties of these various novel device structures.

5.4.6 *Inverted Structures*

The first inverted polymer solar cell, which had the device architecture ITO/Cs_2CO_3/P3HT:PCBM/V_2O_5/Al, was reported in 2006 (Fig. 5.19a). In that study, the influence of the interfacial buffer layer (e.g., Cs_2CO_3, V_2O_5) on the device performance was investigated in detail. The presence of an electrode buffer layer

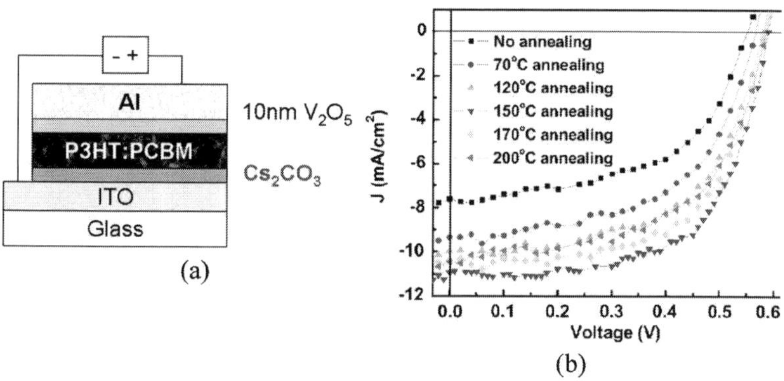

Figure 5.19 (a) Inverted device structure prepared by Liao *et al.* (b) I–V curves (under illumination) of inverted PV devices prepared with annealing of the Cs_2CO_3 layer at various temperatures [54] (Reprinted with permission from Liao, H. H., L. M. Chen, Z. Xu, G. Li, Yang, Y. (2008). *Appl. Phys. Lett.*, **92**, 173303, Copyright 2008, American Institute of Physics).

Table 5.3 Performances of inverted devices and contact angles of Cs_2CO_3 after annealing at various temperatures [54]

Annealing T (°C)	V_{oc} (V)	J_{sc} (mA/cm^2)	PCE (%)	FF (%)	Contact angle (θ,°)	Annealing T (°C)
N/A	0.55	7.61	2.31	55.2	11	N/A
70	0.565	9.36	2.92	55.2	29	70
120	0.58	10.86	3.69	58.6	49	120
150	0.59	11.13	4.19	64	54	150
170	0.582	11.29	4.08	62.1	60	170

drastically enhanced the efficiency of the inverted polymer solar cell from 0.23 to 2.25%, relative to that of the original inverted structure lacking a buffer layer (i.e., ITO/P3HT:PCBM/Al) [53]. Since then, Liao *et al.* have demonstrated several highly efficient inverted polymer solar cells based on the same structure, prepared with thermal annealing treatment of the buffer layer. In particular, thermal annealing at 150°C decreased the work function of Cs_2CO_3 from 3.45 to 3.06 eV, as verified using ultraviolet photoelectron spectroscopy [54]; accordingly, Fig. 5.19b reveals that the PCE improved from 2.31 to 4.19% after thermal annealing of the Cs_2CO_3 interfacial layer at 150°C. To understand these phenomena, contact angle measurements were performed and x-ray photoelectron spectra were recorded for a representative Cs_2CO_3 surface. Table 5.3 reveals that the contact angle of the Cs_2CO_3 surface increased upon increasing the annealing temperature; this transition from a hydrophilic to hydrophobic surface was beneficial for the growth of polymer films, thereby enhancing the device performance. Moreover, the XPS data revealed that thermal annealing caused the Cs_2CO_3 to decompose into low-work-function cesium oxide; this doped cesium oxide behaved as an n-type semiconductor and featured a relatively low work function. Using this approach, the achieved efficiency of 4.2% was comparable with that of a conventional device having a regular structure.

Based on the inverted structure, we recently prepared flexible polymer solar cells on stainless-steel (SS) substrates [55], which provide mechanical flexibility, robustness, and thermal stability.

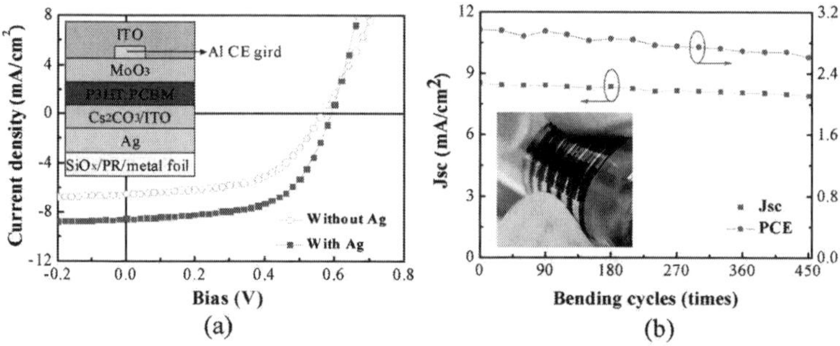

Figure 5.20 (a) $J-V$ curves of flexible OPVs on metal foils, prepared with and without a bottom Ag layer. Inset: Device structure of the flexible OPVs. (b) Plots of J_{sc} and PCE with respect to the number of bending cycles (bending radius: ca. 0.8 cm). Inset: Photograph of the flexible OPVs prepared on a metal foil [55] (Reprinted with permission from Chen, F. C., J. L. Wu, C. L. Lee, W. C. Huang, H. M. P. Chen, Chen, W. C. (2009). Flexible polymer PV devices prepared with inverted structures on metal foils, *IEEE Electron Device Lett.*, **30**, 727–729, Copyright 2009, IEEE).

Compared with plastics, SS substrates can further protect the devices against the penetration of oxygen and moisture from the atmosphere. Nevertheless, the opacity of SS necessitated that these inverted structures were top-illuminated devices. Therefore, for the fabrication of flexible devices on SS substrates, we adopted an inverted structure having the configuration SS foil/photoresist layer/SiO_x/Ag/ITO/Cs_2CO_3/P3HT:PCBM/ MoO_3/Al grid (10%)/ITO (inset to Fig. 5.20a). In these flexible OPVs, the average transmittance in the visible region (400–700 nm) of the transparent top electrode (MoO_3/ITO) was 86.2%, suggesting that photons were transmitted efficiently through the top contact. Figure 5.20a presents the corresponding $J-V$ curves of the devices prepared with and without an Ag layer. The PCE of the flexible OPVs improved from 2.02 to 2.97% after incorporating the functional Ag layer. Notably, the bottom Ag layer reduced the sheet resistance (from 130 to 0.3 Ω/cm), thereby decreasing the series resistance of the device (from 9.1 to 3.3 $\Omega \cdot cm^2$). Moreover, the bottom Ag layer also acted as an effective optical reflector, leading to a considerable enhancement in reflectance at the bottom contact (from ca. 30 to ca. 90% in the visible region), thereby increasing the photocurrent from 6.48 to

8.51 mA/cm^2. To further investigate the mechanical stability of the flexible OPVs, we performed a cyclic bending test at a radius of 0.8 cm for the flexible OPVs made without any further encapsulation. Figure 5.20b reveals that the values of J_{sc} and PCE decreased slightly after 450 bending cycles: from 8.51 to 7.91 mA/cm^2 and from 2.97 to 2.61%, respectively; moreover, the value of V_{oc} remained relatively unchanged (from 0.59 to 0.57 V). Because we performed the stability test under ambient conditions, it is likely that more reliable results would arise after appropriate encapsulation of the flexible devices. These results suggest that inverted structures are excellent candidates for top-illuminated flexible OPVs on metal foils

For inverted devices, the cathodes, which are often made of low-work-function metals (e.g., Ca), are generally embedded deeply. Because the upper thin films behave as natural protection layers of the underlaying cathodes, protecting them against oxygen and moisture from the atmosphere, inverted OPVs usually exhibit superior air stability. For example, Hau *et al.* reported air-stable, low-temperature-processed flexible OPVs having the structure ITO/ZnO/P3HT:PCBM/PEDOT:PSS/Ag (Fig. 5.21a) [56]. Notably, the ZnO NPs provided high electron mobility and high transmittance. Furthermore, because the PEDOT:PSS layer can decrease the degree of oxygen diffusion toward the active layer, the decay of the device was slowed down [57]. In addition, the ZnO was made through a traditional sol–gel process that required further high temperature annealing (400°C), which is not compatible with plastic substrates. Table 5.4 summarizes the device performances of the conventional and inverted solar cells. The conventional (standard) device using LiF and Al as the cathode structure had an average PCE of ca. 2.4%. The device incorporating the high-temperature-processed sol–gel ZnO, however, exhibited a higher PCE of ca. 3.5%. Meanwhile, the devices fabricated with ZnO NPs, which were formed through spin-coating at room temperature on ITO-coated glass substrates, had an average PCE of ca. 3.6%, slightly higher than that of the devices containing ZnO prepared through sol–gel processing at higher temperature. X-ray diffraction data revealed that both the ZnO prepared using sol–gel processing and the ZnO NPs featured typical diffraction peaks at 31.8 (100), 34.3 (002), and 36.3° (101), indicating that ZnO of high crystallinity could be achieved after spin-coating at

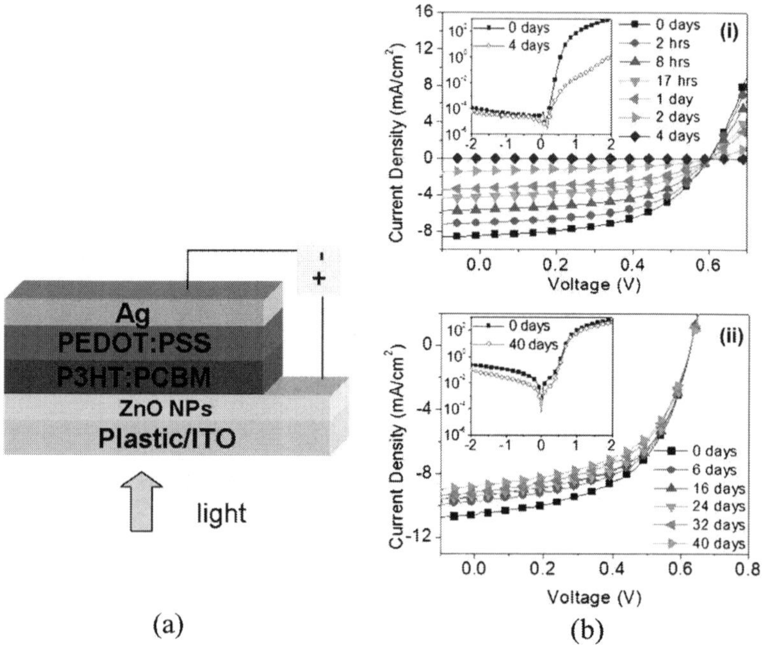

Figure 5.21 (a) Device configuration of the flexible inverted plastic solar cell prepared by Hau *et al.* (b) *J–V* curves of the unencapsulated (i) conventional (over a period of 4 days) and (ii) inverted (over a period of 40 days) devices. Inset: Dark current device characteristics over a period of days (in air under ambient conditions) [56] (Reprinted with permission from Hau, S. K., H. L. Yip, N. S. Baek, J. Y. Zou, O'Malley, K., Jen, A. K. Y. (2008). *Appl. Phys. Lett.*, **92**, 253301 Copyright 2008, American Institute of Physics).

room temperature. The devices fabricated from ZnO NPs on ITO-coated plastics exhibited slightly lower PCEs of ca. 3.3% (Table 5.4), presumably due to the lower transparency of the plastic in the region from 500 to 600 nm, relative to that of the glass substrate. Most importantly, Hau *et al.* demonstrated that environmentally stable inverted polymer solar cells could be prepared on ITO-coated flexible plastic substrates. The device stability of unencapsulated devices featuring the conventional structure and ZnO NPs on ITO-coated plastic substrates was periodically investigated for 40 days under ambient conditions. From the *J–V* curves of the conventional device (Fig. 5.21b), the authors observe a dramatic decrease in

Table 5.4 Device performance of unencapsulated conventional P3HT:PCBM BHJ solar cells incorporating LiF/Al as an electrode and unencapsulated inverted P3HT:PCBM BHJ solar cells incorporating sol–gel-processed ZnO or ZnO NPs on ITO/glass and ZnO NPs on ITO/plastic.

Substrate (electrode)	ZnO	V_{oc} (V)	J_{sc} (mA/cm^2)	FF (%)	PCE (%)
Glass/ITO	N/A	0.598	7.87	50.9	2.4
(LiF/Al)		(0.6)	(8.48)	(51)	(2.6)
Glass/ITO	ZnO sol-gel	0.617	11.08	51.3	3.51
(Ag)		(0.618)	(11.53)	(51.7)	(3.68)
Glass/ITO	ZnO NPs	0.623	10.69	54.2	3.61
(Ag)		(0.623)	(11.17)	(54.3)	(3.78)
Plastic/ITO	ZnO NPs	0.624	9.81	53.9	3.3
(Ag)		(0.629)	(10.6)	(53.5)	(3.58)

Average values were obtained from testing of over 10 devices; numbers in parentheses denote highest performance achieved [56].

Reprinted with permission from Hau, S. K., H. L. Yip, N. S. Baek, J. Y. Zou, O'Malley, K., Jen, A. K. Y. (2008). *Appl. Phys. Lett.*, **92**, 253301. Copyright 2008, American Institute of Physics.

photocurrent after only 4 days, with the dark current at 2 V decreasing by two orders of magnitude relative to its initial state. In contrast, the inverted devices containing the ZnO NPs exhibited superior stability in air. The values of both V_{OC} and FF over a period of 40 days remained relatively constant, while J_{SC} decreased slightly; the PCE remained at over 80% of its initial value. In short, these findings suggest that inverted structures can be used to realize air-stable, low-temperature-processed, flexible polymer solar cells exhibiting high efficiency.

5.4.7 *Stacked Structure*

Stacked structures consisting of multiple photoactive layers are also commonly used for harvesting sunlight more efficiently. In this approach, two polymer PV cells are stacked together and connected in series or in parallel. In 2006, Shrotriya *et al.* demonstrated a stacked structure consisting of two individual polymer solar cells [58]. One of the devices had highly transparent electrodes so that excess photons that the front cell could not harvest would to pass through the transparent device and be absorbed in the back cell. Relative to the corresponding conventional device, the absorption

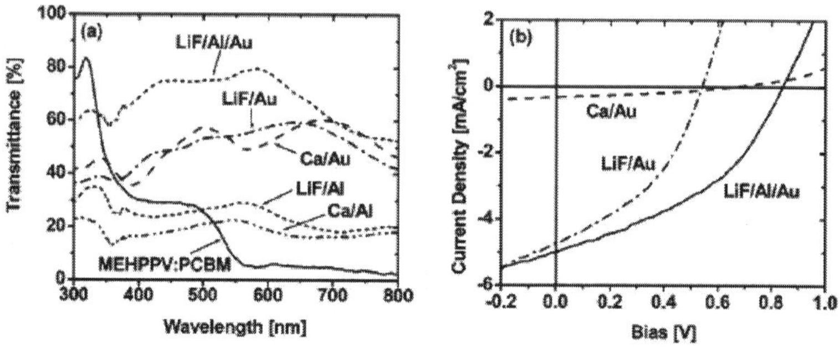

(a) (b)

Figure 5.22 (a) Transmittance spectra of various semitransparent cathodes: Ca (10 nm)/Al(10 nm), Ca (10 nm)/Au (12.5 nm), Ca (10 nm)/Au (12.5 nm), LiF (0.5 nm)/Au (12.5 nm), LiF (1 nm)/Al (2.5 nm)/Au (12.5 nm). The UV–Vis absorption spectrum of an MEH-PPV:PCBM film is provided for comparison. (b) I–V curves for polymer PV cells featuring Ca (10 nm)/Au (12.5 nm), LiF (0.5 nm)/Au (12.5 nm), LiF (1 nm)/Al (2.5 nm)/Au (12.5 nm) as semitransparent cathodes [58].

efficiency improved because the length of the light path had doubled. The photoactive materials used in that study were MEH-PPV and PCBM. To obtain both high transparency and high efficiency for the front semitransparent device, the transparent cathode required high transparency and superior ability to collect electrons. As indicated in Fig. 5.22, semitransparent polymer PVs having a cathode structure LiF (1 nm)/Al (2.5 nm)/Au (12.5 nm) provided the highest device efficiency and the highest transparency. The higher transparency of this cathode structure resulted from three factors: (i) superior wettability between LiF/Al and Au resulted in a continuous film that was free of voids and, therefore, the degree of light scattering was decreased; (ii) Au has a higher skin depth ($\delta = 20$ nm) relative to that of Al; (iii) the thin oxidized Al layer had a high level of transparence.

After the semitransparent device structures were obtained, two separate PV devices could be stacked directly together. The devices were stacked on top of each other and could be connected in series or in parallel to either enhance the voltage or current output, respectively, as indicated in the inset to Fig. 5.23a. When the stacked devices were connected in series, the value of V_{oc} doubled, whereas the values of I_{sc} and FF remained unchanged. On the other hand,

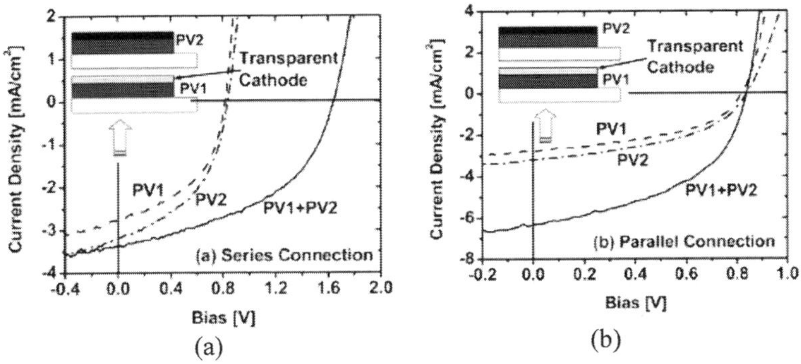

Figure 5.23 I–V curves of multiple-device stacked structures in which the two cells were connected in (a) series and (b) parallel, as well as those of the individually measured PV1 and PV2. Inset: Stacking sequence for these two devices [58] (Reprinted with permission from Shrotriya, V., E. H. E. Wu, G. Li, Y. Yao, Yang, Y. (2006). *Appl. Phys. Lett.*, **88**, 064104, Copyright 2006, American Institute of Physics).

when the two devices were connected in parallel (Fig. 5.23b), the value of J_{SC} doubled, whereas the values of V_{OC} and FF remained unchanged, relative to those of the single-layer devices PV1 and PV2. Overall, the PCEs of the stacked structures were increased to 2.4 ($\pm$0.2)% (in series) and 2.5 ($\pm$0.1)% (in parallel) relative to the values of ca. 1.1 and ca. 1.3% for PV1 and PV2, respectively. These results suggest that stacked structures open a path toward polymer solar cells exhibiting higher efficiencies. Furthermore, this approach can be extended to the use of photoactive materials with different absorption spectra, thereby potentially covering the entire solar spectrum and, as a result, providing superior performance.

5.4.8 *Tandem Structures*

The most remarkable architecture used to obtain highly efficient OPVs has been the tandem structure, which directly connects two single cells in series through an intermediate structure [59–67]. Hiramoto *et al.* reported the first organic tandem cell, prepared after connecting two small molecule–based cells through a semitransparent Au interlayer [65]. In 2002, Yakimov and Forrest stacked two subcells using a layer of Ag nanoclusters, suggesting

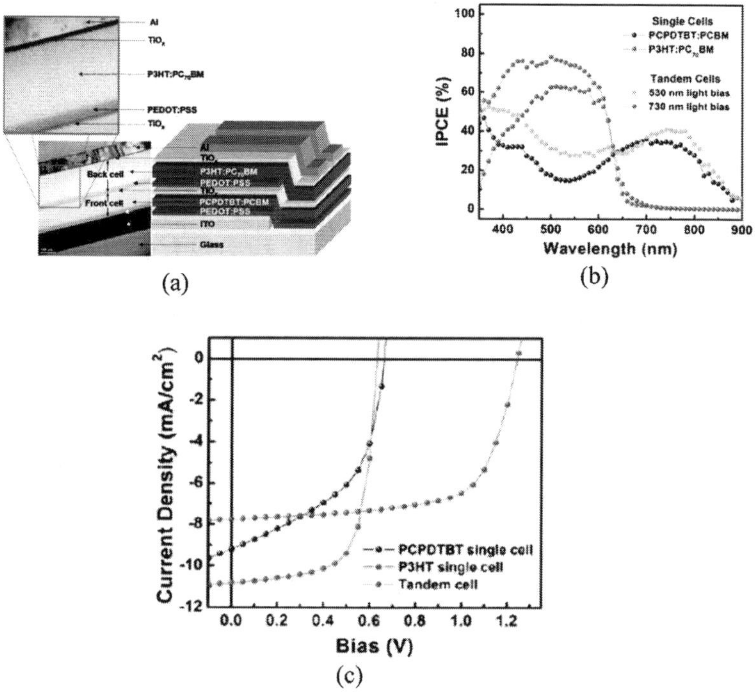

Figure 5.24 (a) Device structure and TEM cross-sectional images of polymer tandem solar cells (lower image: scale bar, 100 nm; upper image: scale bar, 20 nm). (b) Absorption spectra of a PCPDTBT:PCBM film, a $P3HT:PC_{70}BM$ film, and a bilayer of the two in a tandem device structure. (c) J–V characteristics of single cells incorporating PCPDTBT:PCBM and $P3HT:PC_{70}BM$ composites and of a tandem cell incorporating both under AM1.5G illumination from a calibrated solar simulator [irradiation intensity: 100 mW/cm^2 (ca. 1 sun)] [67] (From Kim, J. Y., K. Lee, N. E. Coates, D. Moses, T. Q. Nguyen, M. Dante, Heeger, A. J. (2007), *Science*, **317**, 222–225. Reprinted with permission from AAAS).

that the NPs served as recombination centers [66]. Thereafter, many groups reported tandem cells prepared from polymeric materials. For instance, in 2007 Kim *et al.* used all-solution processing to fabricate tandem organic solar cells, achieving a very high efficiency of 6.5% [67]. Figure 5.24a presents TEM images of the cross-section of the tandem cells. The front cell featured a BHF structure, with a blend of poly[2,6-(4,4-bis-(2-ethylhexyl)-4*H*-cyclopenta[2,1-

b;3,4-b′]dithiophene)-*alt*-4,7-(2,1,3-benzothiadiazole)] (PCPDTBT) and PCBM as the active layer; P3HT and $PC_{70}BM$ were used for the back cell [67]. Meanwhile they used a transparent TiO_x thin film and a PEDOT:PSS layer as intermediate layers to connect the two subcells. Therefore, electrons from the first cell were expected to recombine with holes from the second cell at the TiO_x–PEDOT:PSS interface. The TiO_x layer, which was deposited through sol–gel processing, performed four functions in the device, behaving as an electron transport layer, a hole blocking layer, an optical spacer, and a barrier preventing any washing effect from occurring during the fabrication of the PEDOT:PSS layer. With regard to the absorption spectra of the photoactive layers, the absorption peaks of the PCPDTBT:PCBM and P3HT:$PC_{70}BM$ blends (Fig. 5.24b) sufficiently complemented each other, providing spectral coverage from 300 to 900 nm; i.e., the overall absorption in the tandem cell featured more extended coverage of the solar spectrum. Furthermore, the PCPDTBT:PCBM subcell exhibited values of J_{sc}, V_{oc}, FF, and PCE of 9.2 mA/cm^2, 0.66 V, 0.5, and 3%, respectively; for the P3HT:$PC_{70}BM$ single cell, these values were 10.8 mA/cm^2, 0.63 V, 0.69, and 4.7%, respectively. Superior device efficiency was obtained after combining these two single cells into tandem cells; the values of V_{oc}, J_{sc}, FF, and PCE were 1.24 V, 7.8 mA/cm^2, 0.67, and 6.5%, respectively (Fig. 5.24c). These high-performance parameters indicate that tandem cells fabricated using all-solution processes can provide enhanced coverage of the solar spectrum and, thereby, superior device performance.

5.5 Conclusions and Outlook

Our aim in writing this chapter was to introduce the fundamentals and review the recent literature relating to remarkable approaches used to achieve highly efficient polymer solar cells. We introduced several approaches whereby modification of the electrodes could be used to improve device efficiency and operating stability. We also demonstrated that improvement in the optical properties of polymer solar cells was another alternative approach for device enhancement. For example, the introduction of localized surface plasmons induced by metal NPs, the application of light trapping by nanostructures, and the incorporation of optical spacers to redistribute the electrical field in the device have all been employed successfully. In addition, blending near-infrared dyes in the photoactive layer can be used to extend the spectral coverage. Finally, we discussed the behavior of devices featuring inverted, stacked, and tandem structures. In particular, superior air stability can be achieved from inverted structures; stacked and tandem cells provide higher values of V_{oc} and enlarge the spectral coverage, allowing the collection of more photons from the sunlight.

For single-layer polymer/fullerene solar cells, one theoretical model has predicted that the ultimate PCE would be ca. 11% [68]; in addition, Dennler *et al.* suggested that tandem cells exhibiting PCEs of up to 15% might be feasible [69]. In practice, an impressive PCE of 7.9% was recently announced for a polymer solar cell (a number that has been certified by the U.S. Department of Energy's National Renewable Energy Laboratory) [70]. Thus, polymer solar cells can already complete with other thin-film PV technologies, such as those based on amorphous Si. With the unique advantages provided by plastic solar cells, including flexibility and short energy pay-back times, we expect that the commercialization of high-efficient, large-area polymer solar cells will be realized in the near future.

Acknowledgment

We thank the National Science Council and the Ministry of Education (through ATU program) for financial support.

References

1. Tang, C. W. (1986). Two-layer organic photovoltaic cell. *Appl. Phys. Lett.* **48**, 183–185.
2. Schaper, M., J. Schmidt, H. Plagwitz, and R. Brendel (2005). 20.1%-Efficient crystalline silicon solar cell with amorphous silicon rear-surface passivation. *Prog. Photovoltaics* **13**, 381–386.

3. Shah, A., P. Torres, R. Tscharner, N. Wyrsch, and H. Keppner (1999). Photovoltaic technology: the case for thin-film solar cells. *Science* **285**, 692–698.
4. Baxter, J. B. and Aydil E. S. (2005). Nanowire-based dye-sensitized solar cells. *Appl. Phys. Lett.* **86**, 053114.
5. Contreras, M. A., B. Egaas, K. Ramanathan, J. Hiltner, A. Swartzlander, F. Hasoon, and R. Noufi (1999). Progress toward 20% efficiency in Cu(In,Ca)Se-2 polycrystalline thin-film solar cells. *Prog. Photovoltaics* **7**, 311–316.
6. Chen, L. M., Z. R. Hong, G. Li, and Y. Yang (2009). Recent progress in polymer solar cells: manipulation of polymer: fullerene morphology and the formation of efficient inverted polymer solar cells. *Adv. Mater.* **21**, 1434–1449.
7. Peet, J., M. L. Senatore, A. J. Heeger, and G. C. Bazan (2009). The role of processing in the fabrication and optimization of plastic solar cells. *Adv. Mater.* **21**, 1521–1527.
8. Na, S. I., S. S. Kim, J. Jo, and D. Y. Kim (2008). Efficient and flexible ITO-free organic solar cells using highly conductive polymer anodes. *Adv. Mater.* **20**, 4061–4067.
9. Krebs, F. C. (2009). Polymer solar cell modules prepared using roll-to-roll methods: Knife-over-edge coating, slot-die coating and screen printing. *Sol. Energy Mater. Sol. Cells* **93**, 465–475.
10. Krebs, F. C., J. Alstrup, H. Spanggaard, K. Larsen, and E. Kold (2004). Production of large-area polymer solar cells by industrial silk screen printing, lifetime considerations and lamination with polyethyleneterephthalate. *Sol. Energy Mater. Sol. Cells* **83**, 293–300.
11. Al-Ibrahim, M., H. K. Roth, and S. Sensfuss (2004). Efficient large-area polymer solar cells on flexible substrates. *Appl. Phys. Lett.* **85**, 1481–1483.
12. Hoth, C. N., S. A. Choulis, P. Schilinsky, and C. J. Brabec (2007). High photovoltaic performance of inkjet printed polymer: fullerene blends. *Adv. Mater.* **19**, 3973–3978.
13. De Bettignies, R., J. Leroy, M. Firon, and C. Sentein (2006). Accelerated lifetime measurements of P3HT:PCBM solar cells. *Synth. Met.* **156**, 510–513.
14. Lloyd, M. T., D. C. Olson, P. Lu, E. Fang, D. L. Moore, M. S. White, M. O. Reese, D. S. Ginley and J. W. P. Hsu (2009). Impact of contact evolution on the shelf life of organic solar cells. *J. Mater. Chem.* **19**, 7638–7642.

15. Xue, J. G., B. P. Rand, S. Uchida, and S. R. Forrest (2005). A hybrid planar-mixed molecular heterojunction photovoltaic cell. *Adv. Mater.* **17**, 66–71.
16. Liang, Y. Y., Y. Wu, D. Q. Feng, S. T. Tsai, H. J. Son, G. Li, and L. P. Yu (2009). Development of new semiconducting polymers for high performance solar cells. *J. Am. Chem. Soc.* **131**, 56–57.
17. Park, S. H., A. Roy, S. Beaupre, S. Cho, N. Coates, J. S. Moon, D. Moses, M. Leclerc, K. Lee, and A. J. Heeger (2009). Bulk heterojunction solar cells with internal quantum efficiency approaching 100%. *Nat. Photonics* **3**, 297–303.
18. Krebs, F. C. and H. Spanggaard (2005). Significant improvement of polymer solar cell stability. *Chem. Mater.* **17**, 5235–5237.
19. Ma, W. L., C. Y. Yang, X. Gong, K. Lee, and A. J. Heeger (2005). Thermally stable, efficient polymer solar cells with nanoscale control of the interpenetrating network morphology. *Adv. Funct. Mater.* **15**, 1617–1622.
20. Peet, J., J. Y. Kim, N. E. Coates, W. L. Ma, D. Moses, A. J. Heeger, and G. C. Bazan (2007). Efficiency enhancement in low-bandgap polymer solar cells by processing with alkane dithiols. *Nat. Mater.* **6**, 497–500.
21. Moliton, A., and J. M. Nunzi (2006) How to model the behaviour of organic photovoltaic cells. *Polym. Int.* **55**, 583.
22. Waldauf, C., M. C. Scharber, P. Schilinsky, J. A. Hauch, and C. J. Brabec (2006). Physics of organic bulk heterojunction devices for photovoltaic applications. *J. Appl. Phys.* **99**, 104503.
23. Malliaras, G. G., J. R. Salem, P. J. Brock, and J. C. Scott (1998). Photovoltaic measurement of the built-in potential in organic light emitting diodes and photodiodes. *J. Appl. Phys.* **84**, 1583–1587.
24. Peumans, P., A. Yakimov, and S. R. Forrest (2003). Small molecular weight organic thin-film photodetectors and solar cells. *J. Appl. Phys.* **93**, 3693–3723.
25. Kumar, P., S. C. Jain, V. Kumar, S. Chand, and R. P. Tandon (2009). A model for the $J-V$ characteristics of P3HT:PCBM solar cells. *J. Appl. Phys.* **105**, 104507.
26. Park, B., and M. Y. Han (2009). Photovoltaic characteristics of polymer solar cells fabricated by pre-metered coating process. *Opt. Express* **17**, 13830–13840.
27. Shaheen, S. E., and R. Radspinner, N. Peyghambarian, G. E. Jabbour (2001). Fabrication of bulk heterojunction plastic solar cells by screen printing. *Appl. Phys. Lett.* **79**, 2996–2998.

28. Kim, S. S., S. I. Na, J. Jo, G. Tae, and D. Y. Kim (2007). Efficient polymer solar cells fabricated by simple brush painting. *Adv. Mater.* **19**, 4410–4415.

29. Gunes, S., H. Neugebauer, and N. S. Sariciftci (2007). Conjugated polymer-based organic solar cells. *Chem. Rev.* **107**, 1324–1338.

30. Yu, G., J. Gao, J. C. Hummelen, F. Wudl, and A. J. Heeger (1995). Polymer photovoltaic cells—enhanced efficiencies via a network of internal donor-acceptor heterojunctions. *Science* **270**, 1789–1791.

31. Schilinsky, P., C. Waldauf, and C. J. Brabec (2002). Recombination and loss analysis in polythiophene based bulk heterojunction photodetectors. *Appl. Phys. Lett.* **81**, 3885–3887.

32. Wudl, F. (1992). The chemical-properties of buckminsterfullerene (C_{60}) and the birth and infancy of fulleroids. *Acc. Chem. Res.* **25**, 157–161.

33. Li, G., V. Shrotriya, J. S. Huang, Y. Yao, T. Moriarty, K. Emery, and Y. Yang (2005). High-efficiency solution processable polymer photovoltaic cells by self-organization of polymer blends. *Nat. Mater.* **4**, 864–868.

34. Padinger, F., R. S. Rittberger, and N. S. Sariciftci (2003). Effects of postproduction treatment on plastic solar cells. *Adv. Funct. Mater.* **13**, 85–88.

35. Reyes-Reyes, M., K. Kim, and D. L. Carroll (2005). High-efficiency photovoltaic devices based on annealed poly(3-hexylthiophene) and 1-(3-methoxycarbonyl)-propyl-1-phenyl-(6,6)C_{61} blends. *Appl. Phys. Lett.* **87**, 083506.

36. Zhao, Y., Z. Y. Xie, Y. Qu, Y. H. Geng, and L. X. Wang (2007). Solvent-vapor treatment induced performance enhancement of poly(3-hexylthiophene):methanofullerene bulk-heterojunction photovoltaic cells. *Appl. Phys. Lett.* **90**, 043504.

37. Ko, C. J., Y. K. Lin, and F. C. Chen (2007). Microwave annealing of polymer photovoltaic devices. *Adv. Mater.* **19**, 3520–3523.

38. Chen, F. C., H. C. Tseng, and C. J. Ko (2008). Solvent mixtures for improving device efficiency of polymer photovoltaic devices. *Appl. Phys. Lett.* **92**, 103306.

39. Ko, C. J., Y. K. Lin, F. C. Chen, and C. W. Chu (2007). Modified buffer layers for polymer photovoltaic devices. *Appl. Phys. Lett.* **90**, 063509.

40. Yip, H. L., S. K. Hau, N. S. Baek A. K., and Y. Jen (2008). Self-assembled monolayer modified ZnO/metal bilayer cathodes for polymer/fullerene bulk-heterojunction solar cells. *Appl. Phys. Lett.* **92**, 193313.

41. Chen, F. C., and S. C. Chien (2009). Nanoscale functional interlayers formed through spontaneous vertical phase separation in polymer photovoltaic devices. *J. Mater. Chem.* **19**, 6865–6869.
42. Morfa, A. J., K. L. Rowlen, T. H. Reilly, M. J. and Romero van de Lagemaat, J. (2008). Plasmon-enhanced solar energy conversion in organic bulk heterojunction photovoltaics. *Appl. Phys. Lett.* **92**, 013504.
43. Kim, S. S., S. I. Na, J. Jo, D. Y. Kim, and Y. C. Nah (2008). Plasmon enhanced performance of organic solar cells using electrodeposited Ag nanoparticles. *Appl. Phys. Lett.* **93**, 073307.
44. Lee, J. H., J. H. Park, J. S. Kim, D. Y. Lee, and K. Cho (2009). High efficiency polymer solar cells with wet deposited plasmonic gold nanodots. *Org. Electron.* **10**, 416–420.
45. Chen, F. C., J. L. Wu, C. L. Lee, Y. Hong, C. H. Kuo, and M. H. Huang (2009). Plasmonic-enhanced polymer photovoltaic devices incorporating solution-processable metal nanoparticles. *Appl. Phys. Lett.* **95**, 013305.
46. Mihailetchi, V. D., L. J. A. Koster, J. C. Hummelen, and P. W. M. Blom (2004). Photocurrent generation in polymer-fullerene bulk heterojunctions. *Phys. Rev. Lett.* **93**, 216601.
47. Mihailetchi, V. D., H. X. Xie de Boer, B., L. J. A. Koster, and P. W. M. Blom (2006). Charge transport and photocurrent generation in poly(3-hexylthiophene):methanofullerene bulk-heterojunction solar cells. *Adv. Funct. Mater.* **16**, 699–708.
48. Na, S. I., Seok-Soon, K., S. S. Kwon, J. Jang, K. Juhwan, T. Lee, and Dong-Yu, K. (2007). Surface relief gratings on poly(3-hexylthiophene) and fullerene blends for efficient organic solar cells. *Appl. Phys. Lett.* **91**, 073509.
49. Snaith, H. J., N. C. Greenham, and R. H. Friend (2004). The origin of collected charge and open-circuit voltage in blended polyfluorene photovoltaic devices. *Adv. Mater.* **16**, 1640ff.
50. Kim, J. Y., S. H. Kim, H. H. Lee, K. Lee, W. L. Ma, X. Gong, and A. J. Heeger (2006). New architecture for high-efficiency polymer photovoltaic cells using solution-based titanium oxide as an optical spacer. *Adv. Mater.* **18**, 572–576.
51. Gilot, J., I. Barbu, M. M. Wienk, and R. A. J. Janssen (2007). The use of ZnO as optical spacer in polymer solar cells: Theoretical and experimental study. *Appl. Phys. Lett.* **91**, 113520.
52. Peet, J., A. B. Tamayo, X. D. Dang, J. H. Seo, and T. Q. Nguyena (2008). Small molecule sensitizers for near-infrared absorption in polymer bulk heterojunction solar cells. *Appl. Phys. Lett.* **93**, 163306.

53. Li, G., C. W. Chu, V. Shrotriya, J. Huang, and Y. Yang (2006). Efficient inverted polymer solar cells. *Appl. Phys. Lett.* **88**, 253503.
54. Liao, H. H., L. M. Chen, Z. Xu, G. Li, and Y. Yang (2008). Highly efficient inverted polymer solar cell by low temperature annealing of Cs_2CO_3 interlayer. *Appl. Phys. Lett.* **92**, 173303.
55. Chen, F. C., J. L. Wu, C. L. Lee, W. C. Huang H. M. P. Chen, and W. C. Chen (2009). Flexible polymer photovoltaic devices prepared with inverted structures on metal foils. *IEEE Electron. Device. Lett.* **30**, 727–729.
56. Hau, S. K., H. L. Yip, N. S. Baek, J. Y. Zou O'Malley, K., and A. K. Y. Jen (2008). Air-stable inverted flexible polymer solar cells using zinc oxide nanoparticles as an electron selective layer. *Appl. Phys. Lett.* **92**, 253301.
57. Andersen, M., J. E. Carle Cruys-Bagger, N., M. R. Lilliedal, and M. A. Hammond Winther-Jensen, B, F. C. Krebs (2007). Transparent anodes for polymer photovoltaics: Oxygen permeability of PEDOT. *Sol. Energy Mater. Sol. Cells* **91**, 539–543.
58. Shrotriya, V., E. H. E. Wu, G. Li, Y. Yao, and Y. Yang (2006). Efficient light harvesting in multiple-device stacked structure for polymer solar cells. *Appl. Phys. Lett.* **88**, 064104.
59. Kim, C. and J. Kim (2008). Organic photovoltaic cell in lateral-tandem configuration employing continuously-tuned microcavity sub-cells. *Opt. Express* **16**, 19987–19994.
60. Zhao, D. W., X. W. Sun, C. Y. Jiang A. K. K. Kyaw, G. Q. Lo, and D. L. Kwong (2008). Efficient tandem organic solar cells with an Al/MoO_3 intermediate layer. *Appl. Phys. Lett.* **93**, 083305.
61. Zhang, C. F., S. W. Tong, C. Y. Jiang, E. T. Kang D. S. H. Chan, and C. X. Zhu (2008). Simple tandem organic photovoltaic cells for improved energy conversion efficiency. *Appl. Phys. Lett.* **92**, 083310.
62. Tvingstedt, K., V. Andersson, F. Zhang, and O. Inganas (2007). Folded reflective tandem polymer solar cell doubles efficiency. *Appl. Phys. Lett.* **91**, 123514.
63. Hadipour, A., de Boer, B., and P. W. M. Blom (2008). Organic tandem and multijunction solar cells. *Adv. Funct. Mater.* **18**, 169–181.
64. Chen, F. C. and Lin., C. H (2009). Construction and characteristics of tandem organic solar cells featuring small molecule-based films on polymer-based subcells. *J. Phys. D: Appl. Phys* **43**, 025104.
65. Hiramoto, M., M. Suezaki, and M. Yokoyama (1990). Effect of thin gold interstitial-layer on the photovoltaic properties of tandem organic solar-cell. *Chem. Lett.*, 327–330.

66. Yakimov, A. and S. R. Forrest (2002). High photovoltage multiple-heterojunction organic solar cells incorporating interfacial metallic nanoclusters. *Appl. Phys. Lett.* **80**, 1667–1669.
67. Kim, J. Y., K. Lee, N. E. Coates, D. Moses, T. Q. Nguyen, M. Dante, and A. J. Heeger (2007). Efficient tandem polymer solar cells fabricated by all-solution processing. *Science* **317**, 222–225.
68. Koster, L. J. A., V. D. Mihailetchi, and P. W. M. Blom (2006). Ultimate efficiency of polymer/fullerene bulk heterojunction solar cells. *Appl. Phys. Lett.* **88**, 093511.
69. Dennler, G., M. C. Scharber, T. Ameri, P. Denk, K. Forberich, C. Waldauf, and C. J. Brabec (2008). Design rules for donors in bulk-heterojunction tandem solar cells-towards 15% energy-conversion efficiency. *Adv. Mater.* **20**, 579ff.
70. http://www.solarmer.com/newsevents.php.

Chapter 6

Applications

Yu-Chiang Chao
Institute of Physics, National Chiao Tung University, Hsinchu, Taiwan
yuchiangchao@gmail.com

Polymer-based devices have been finding their way into a growing number of applications, resulting from ongoing improvements in lifespan and performance. Innovations at both the conceptual and practical level have provided opportunities for polymer electronic devices to be used in a wide range of emerging fields. In this chapter, three representative devices, integrated with various polymer electronic devices, are described. First, the idea of polymer light-emitting transistors is introduced, in which an organic light-emitting diode is vertically integrated with a vertical transistor to modulate driving current, and in so doing, modulate luminescence. The second part of this chapter focuses on a pressure sensor with low operational voltage, in which rubber is layered on a vertical transistor, suitable for application as artificial skin. Finally, polymer infrared proximity sensors are described, in which a near infrared light source is integrated with a photo detector to detect the movement of nearby objects.

Polymer Electronics
Edited by Hsin-Fei Meng

ISBN 978-981-4267-84-7 (Hardcover), 978-981-4364-04-1 (eBook)
www.panstanford.com

6.1 Light-Emitting Transistors

Organic light-emitting diode (OLED) displays are thin, light weight, and energy efficient. OLED displays can be produced as thin as a pane of glass, which is far thinner and lighter than existing plasma display panels or liquid crystal displays. Eventually, as the fabrication processes mature, and larger scale manufacturing capacity becomes a reality, OLED displays will begin appearing on the market at a reasonable price.

OLED displays are usually addressed by active-matrix transistor arrays, controlling the current to each pixel [1–4]. The pixels in the liquid crystal display can be voltage switched using a single switching transistor; however, the current-controlled emission of an OLED requires at least two transistors to maintain a continuous current for the duration of the frame [1, 4]. Many driving schemes have been proposed to improve performance [3]. The basic driving scheme of an active-matrix organic electroluminescence (EL) display device comprises a capacitor, an organic EL element, a switching transistor, and a driving transistor, as shown in Fig. 6.1. For this reason, the pixel circuits of active matrix OLED displays occupy more space, often resulting in a much lower aperture ratio than liquid crystal displays. The brightness of a pixel from an OLED is controlled by the current to the pixel, which is modulated by the gate-to-source voltage of the driving transistor. The output current

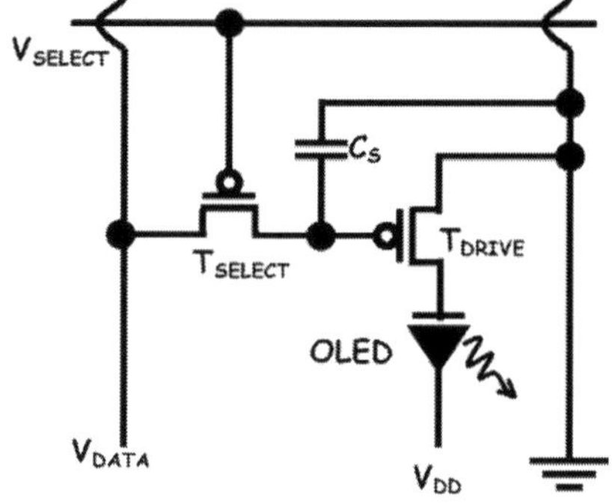

Figure 6.1 Pixel circuit for active-matrix OLED [1].

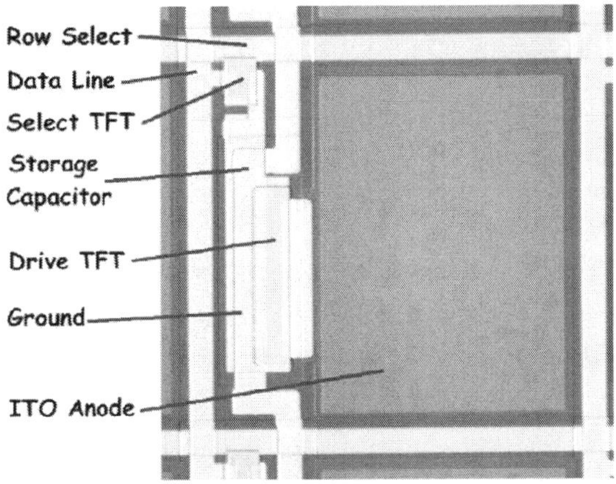

Figure 6.2 Microscopic image of active-matrix OLED pixel [1].

of the driving transistor must be uniform and very stable to maintain constant brightness. Currently, organic field-effect transistors have rather low carrier mobility and problems with stability over the long term [5]. In addition, field-effect transistors are horizontal devices, while Organic light-emitting diodes (PLEDs) are vertical devices, allowing them to be placed side by side, as shown in Fig. 6.2. Unfortunately, the aperture ratio of the pixel is limited by the size of the field-effect transistors. OLEDs can also be stacked on top of the transistor array via a hole for the electrical connection [6], as shown in Fig. 6.3. Nonetheless, an increased number of fabrication steps are required to accommodate this approach. Clearly, the development of alternate structures is essential for the further development of such devices.

6.1.1 *Organic Light-Emitting Field-Effect Transistors*

Organic light-emitting field-effect transistors [7–10], combining electrical switching and the light generation capability in a single device, were developed to replace conventional OLEDs driven by field-effect transistors. Usually, light-emitting field-effect transistors have the same structure as field-effect transistors, (Fig. 6.4a). Because the planar structure enables direct imaging of the EL, organic light-emitting field-effect transistors are attractive devices

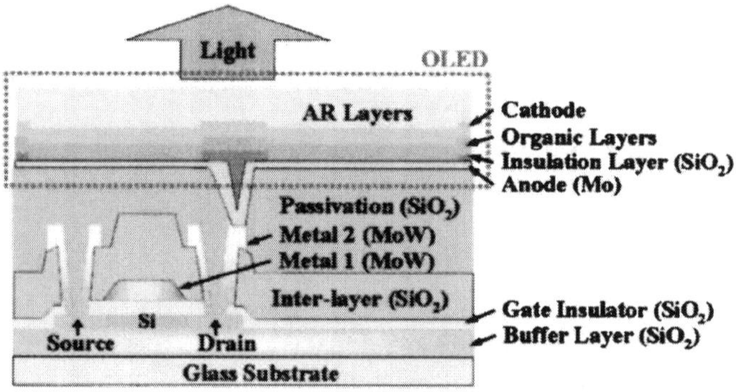

Figure 6.3 Integration of OLED with top-gate low-temperature polycrystalline silicon thin-film transistor [6].

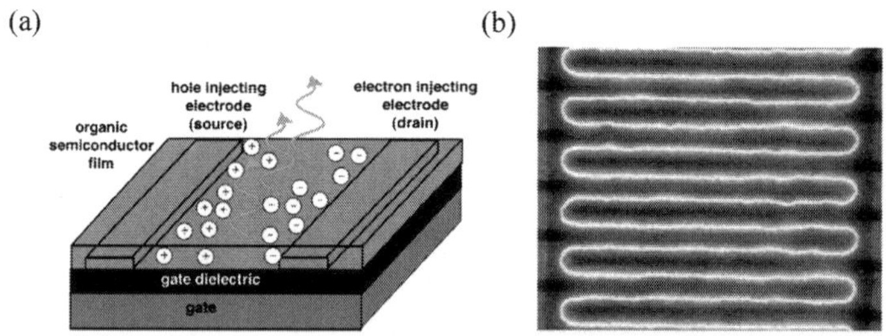

Figure 6.4 (a) Device structure of an organic light-emitting field-effect transistor [7]; (b) Optical microscope image of an operating organic light-emitting transistor [10].

for fundamental researches into carrier injection, transport, and recombination in organic semiconductors. Efficient injection and combination of electrons and holes is crucial for these devices. Figure 6.4b shows an optical microscopic image of an operating light-emitting transistor [10]. Light emission can be observed within the channel at fixed position when applying constant gate voltage and source–drain voltage, and the zone from which light is emitted shifts as the gate voltage is varied. However, so far, the application of light-emitting transistors has been hindered due to low emission efficiency, and the fact that the emission zone is a narrow line pattern, is unfavorable for display applications. The operational voltage, usually exceeding 40 volts, is also too high for real-world applications.

6.1.2 *Polymer Light-Emitting Vertical Transistors*

Light-emitting transistors have been produced by vertically integrating vertical transistors and polymer light-emitting diodes [11]. This vertical integration can be achieved by either fabricating PLEDs on the vertical transistor or fabricating vertical transistors on OLEDs, as shown in Fig. 6.5. The light emitted from the PLED is not shielded by the circuit, and the transistor is used to modulate the current, thereby modulating the luminescence of the PLED. When the PLED is fabricated on a vertical transistor, the light is emitted through the top, transparent cathode, as shown in Fig. 6.5a. When the vertical transistor is fabricated on a PLED, the light is emitted through the bottom transparent cathode as shown in Fig. 6.5b. The top anode can be any opaque metal with high work function. The bottom transparent cathode can be achieved by adopting the cathode used in inverted bottom-emission PLED.

6.1.2.1 Vertical transistor: polymer space-charge-limited transistor

A top-emitting light-emitting vertical transistor has been demonstrated with a vertically integrated top-emitting polymer light-emitting diode on a polymer space-charge-limited transistor (SCLT)

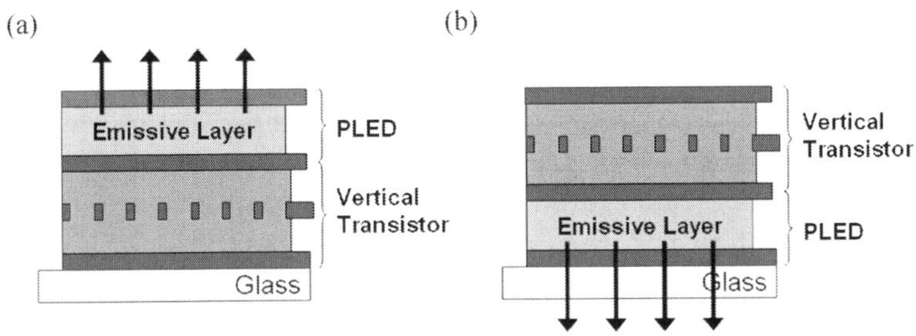

Figure 6.5 Schematic diagram of structure of devices (a) top-emitting light-emitting vertical transistor; and (b) bottom-emitting light-emitting vertical transistor.

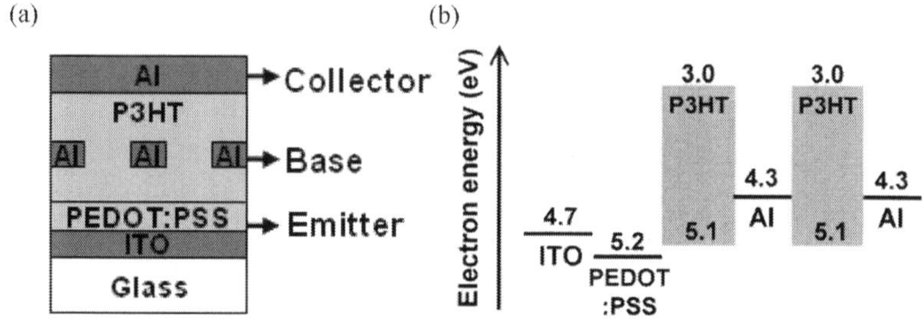

Figure 6.6 (a) Device structure of a polymer space-charge-limited transistor; (b) Diagram of energy band of polymer space-charge-limited transistor.

[11]. The SCLT functions like a switch, modulating the current, and hence the luminance of the polymer light-emitting diode. A schematic illustration of the active area within the structure of the SCLT is shown in Fig. 6.6a. An energy band diagram of the SCLT is shown in Fig. 6.6b. The highest occupied molecular orbital (HOMO) and lowest unoccupied molecular orbital (LUMO) energy for poly(3-hexylthiophene) (P3HT) are 3 and 5.1 eV, respectively. Poly(3,4-ethylenedioxythiophene):poly(styrenesulfonate) (PEDOT: PSS) with a work function of 5.2 eV was used as the emitter for hole injection into the HOMO of P3HT. Al, with a work function of 4.3 eV was used as base and collector. The base electrode was a grid comprising a dense collection of self-assembled submicron openings. The holes are injected from the emitter into the semiconductor, passing through the openings within the metal grid base, before arriving at the collector. The potential distribution and current flow between the emitter and collector is controlled by the voltages of the grid and collector. The potential profile for the *on* and *off* states of the SCLT is shown in Fig. 6.7. The dashed line follows the channel through the grid opening, while the solid line represents the Fermi level of Al and PEDOT. When voltages applied to the grid and the collector constitute a high barrier (Fig. 6.7b) between the emitter and the collector, few holes arrive at the collector through the openings, and hence the SCLT is in the *off* state. Without a barrier (Fig. 6.7a) between the emitter and collector, the holes can pass through the openings to reach the collector, and hence the SCLT is in the *on* state. Organic field-effect transistors are easily influenced by

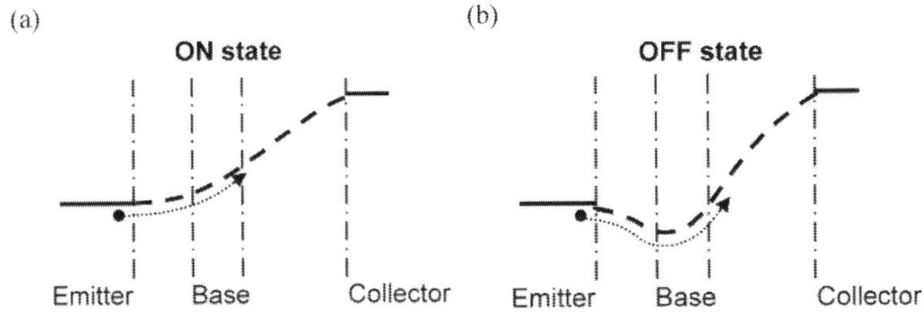

Figure 6.7 Potential profile for the (a) *on* and (b) *off* states of the SCLT.

chemical or physical defects at the gate dielectric-semiconductor interface; however, SCLTs operate independently of the conditions at the dielectric interface, because the current is transported within the bulk.

The structure and principles on which the polymer SCLT operates resemble those of vacuum tube triodes. Vacuum tube triodes consist of a cathode for electron emissions through heating, an anode for electron collection, and a grid for current modulation. Both the grid and anode electrodes are capable of controlling the potential within the device, but the grid is much more effective in controlling the gradient near the cathode. When the electrons experience a negative gradient of potential after being emitted from the cathode, they encounter a large energy barrier, preventing them from being collected by the anode. Conversely, if the electrons encounter a negligible or no energy barrier, many of the electrons can be collected by the anode. However, vacuum tube triodes and SCLTs still differ in two basic aspects. First, in a vacuum tube triode, the charge carriers are electrons; while in SCLTs, the charge carriers are holes. Second, in vacuum tube triodes, the electron is injected into a vacuum from a heated cathode through thermionic emission; while in SCLTs, the hole is injected into HOMO of semiconducting polymer via Ohmic contact.

The fabrication process of SCLTs is shown in Fig. 6.8. A layer of PEDOT:PSS and a layer of P3HT are prepared on ITO as the hole injection layer and semiconducting layer. Polystyrene spheres are adsorbed onto the P3HT surface to serve as a shadow mask, and these spheres are removed after the deposition of Al as a grid.

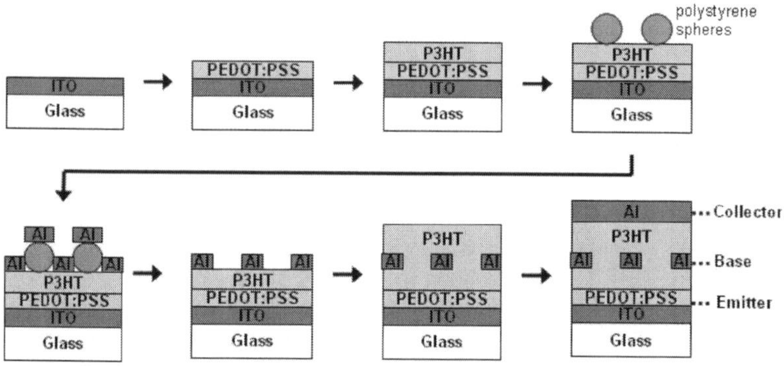

Figure 6.8 Fabrication procedures of the polymer space-charge-limited transistor.

Another layer of P3HT is then deposited, followed by deposit of Al to serve as a collector.

An Al grid with submicron openings has been fabricated using polystyrene spheres as a shadow mask [12]. The polystyrene spheres were first spin coated onto a semiconducting polymer, whereupon Al was deposited to form an Al grid. Figure 6.9 shows

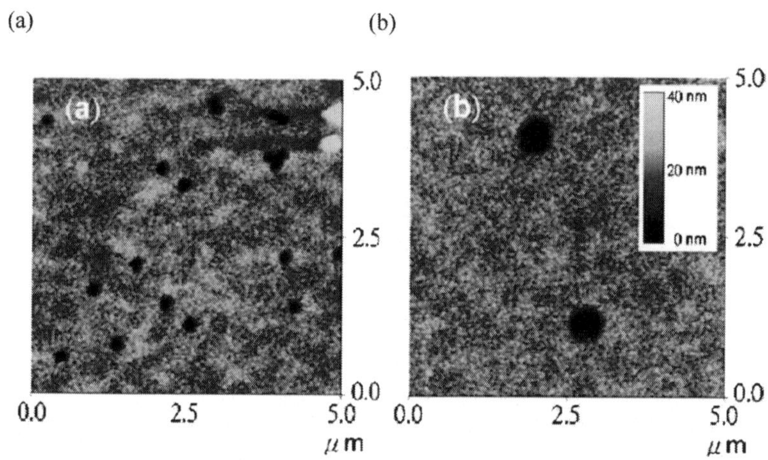

Figure 6.9 Atomic force microscopic images of the Al grid with opening diameters of (a) 200 nm; and (b) 500 nm; the openings were formed in the position of the polystyrene spheres [12].

atomic force microscopic images of the Al grid following removal of the polystyrene spheres [12]. The diameters of the spheres, used as a shadow mask, were 200 nm and 500 nm, and the openings in the grid are formed at the position of polystyrene spheres. The size of the openings was nearly identical to the diameter of the polystyrene spheres, which meant that the removal of the polystyrene spheres did not cause mechanical damage to the Al grid. To prevent the polystyrene spheres from aggregating, the concentration of polystyrene spheres in the solution had to be low, while the speed of the spin coating process had to be high. This approach prevented aggregation of the spheres, but the low density of the polystyrene spheres diminished the density of openings in the Al grid.

By using the spin coating method to form polystyrene spheres on P3HT, SCLT has been realized [12]. Figures 6.10a and 6.10b show the structure of the device and the characteristics of the polymer SCLT with openings on an Al grid with a diameter of 200 nm. The Al collectors was negatively biased at V_{CE} while keeping the emitter commonly grounded. The output current density (J_C) of the collector showed apparent modulation in accordance with the grid voltage V_{GE}. The grid current density was in the order of 10^{-5} mA/cm^2 when various V_{GE} were applied. Figure 6.11 shows the polymer SCLT with openings in the Al grid of 100 nm and 500 nm in diameter. It demonstrated similar characteristics to those in Fig. 6.10b. Turning off the current was more difficult for SCLT with larger openings because the region at the center of the opening was too

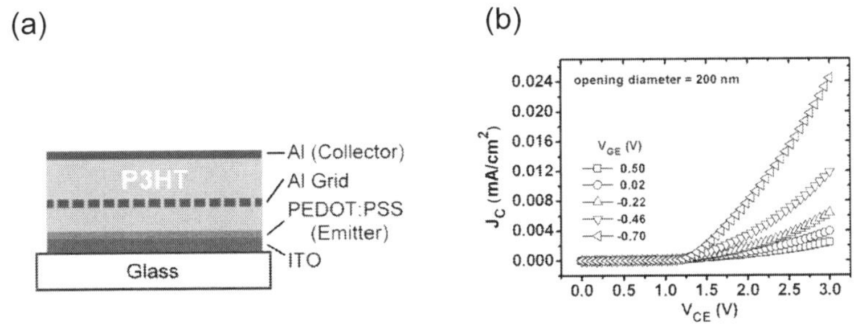

Figure 6.10 (a) Schematic showing the structure of the SCLT; (b) Collector current density J_C as a function of collector voltage V_{CE} at different grid bias V_{GE} [12].

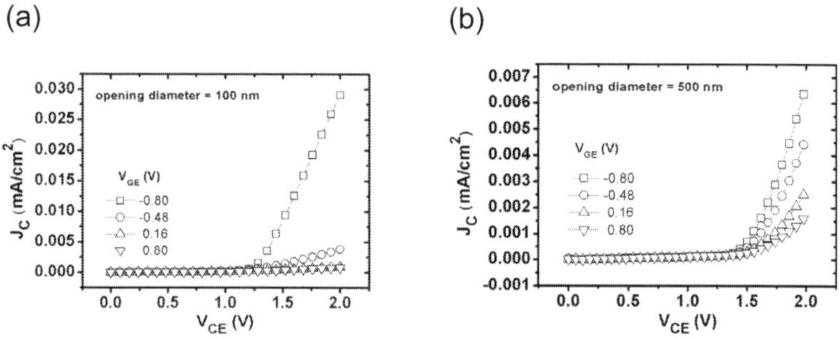

Figure 6.11 Device characteristics of the SCLT with opening diameter of (a) 100 nm; and (b) 500 nm on Al grid [12].

far from the Al grid for the potential barrier to be enhanced more than slightly by the grid potential. All of these devices showed output current density below 10^{-2} mA/cm^2, which would be too low for applications. The on/off ratio below 100 was also too low.

To increase the output current density, a new method has been developed to deposit high-density polystyrene spheres on semiconducting polymers [13]. An ethanol solution with uniformly dispersed polystyrene spheres was first prepared. The substrate was coated with semiconducting polymer before being submerged in the polystyrene solution for tens of seconds. While the substrate was submerged in the solution, the charged polystyrene spheres were adsorbed on the surface of the P3HT without connecting to each other, due to the electrostatic repulsion force. The key step of preventing aggregation of the polystyrene spheres was the immediately transfer of the substrate into a beaker filled with a solution of boiling isopropanol [13]. Atomic force microscopic images of the Al grid with openings 100 nm and 200 nm in diameter are shown in Fig. 6.12. The density of the openings on the Al grid indeed increased and the openings remained separated following treatment of the substrate with boiling isopropanol. The density of openings in the Al grid increased with an increase in the concentration of the solution and submerging time. Despite an increase in the number of openings of irregular shape and size using this process, under optimized conditions, the maximum density of openings with the minimum number of unwanted irregular

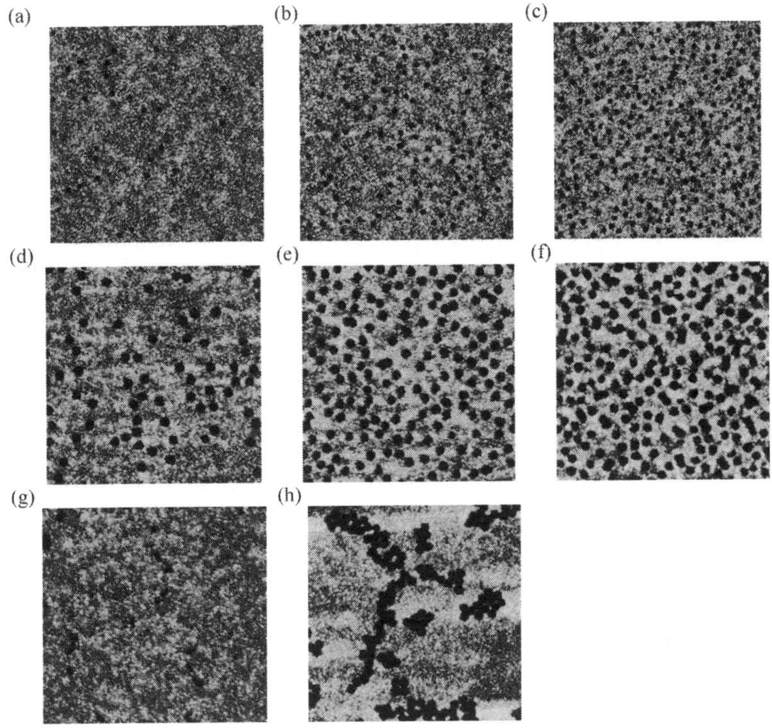

Figure 6.12 Atomic force microscopy images (5 × 5 μm^2), of the surface of Al grid. The height scale is 30 nm. Al grid with opening diameter of 100 nm prepared by submerging substrate in (a) 0.01 wt%; (b) 0.05 wt%; and (c) 0.1 wt% polystyrene solutions for 20 s using boiling isopropanol treatment. Al grid with opening diameter of 200 nm prepared by submerging the substrate in (d) 0.08 wt%; (e) 0.24 wt%; and (f) 0.4 wt% polystyrene solutions for 40 s using boiling isopropanol treatment; (g) Same conditions as (c), but without boiling isopropanol treatment; (h) Same conditions as (f) but without boiling isopropanol treatment [13].

openings could still be achieved by tuning the concentration of the solution and submerging time. However, as shown in Figs. 6.12g and h, nonuniform and connected distribution of spheres within the Al grid formed if the boiling isopropanol treatment was skipped. Nevertheless, an SCLT with high output current density was realized by using this fabrication process, which increases the density of the openings on the Al grid. The transistor characteristics of SCLT with

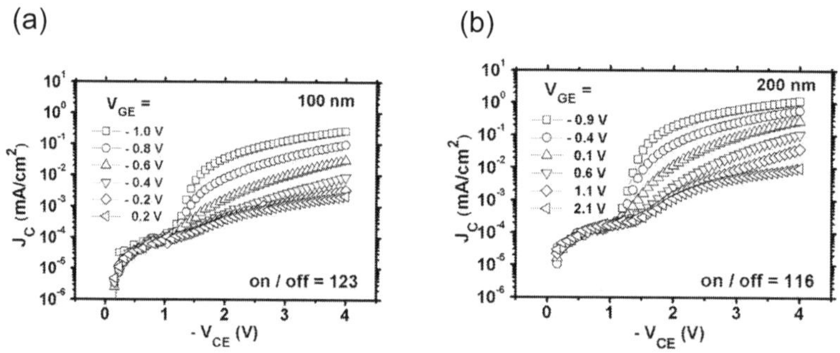

Figure 6.13 Characteristics of transistors of polymer SCLT; (a) Device with grid openings of 100 nm; (b) Device with grid openings of 200 nm [13].

opening diameters of 100 nm and 200 nm are shown in Fig. 6.13. For SCLTs with openings 100 nm and 200 nm in diameter, the on/off ratio was 123 and 116, respectively. The grid current density (J_G) of these devices was in the order of 10^{-4} mA/cm^2, which was much smaller than J_C and thus, the current gain J_C/J_G could be as large as 10^4. With an increase in the density of openings, the output current density of SCLT increased to 0.26 mA/cm^2 for 100 nm device and 1.14 mA/cm^2 for 200 nm devices. The current density was raised by two orders of magnitude, which was nearly enough to drive a high-efficiency phosphorescent organic LED of the same area.

A further increase in output current density in the SCLT was achieved through thermal annealing in a high vacuum, facilitating polymer self-organization into two-dimensional sheets by means of interchain stacking [11]. Self-organization has been shown to improve carrier mobility in P3HT by a few orders of magnitude. In addition, in the SCLT with 200 nm grid openings, an insulating SiO layer was deposited on the Al grid to reduce current leakage from the Al grid to the collector. As shown in Fig. 6.14, using the SiO layer, the output current density of the thermal annealed SCLT could be 27 mA/cm^2 when V_C is -7 V and V_G is -0.9 V. The on–off ratio was 386 when V_C was -7 V. The maximum on/off ratio was 428 when V_C was -6 V. Such output current density was approximately the same as PLED under normal operation. In this manner, the transistor was able to drive a vertically stacked PLED of roughly the same area [11].

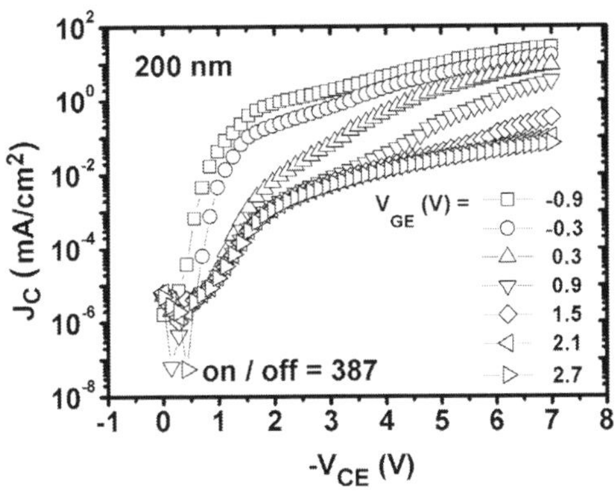

Figure 6.14 Characteristics of transistors of polymer SCLT with thermal annealing treatment in high vacuum. The diameter of the openings in the Al grid is 200 nm [11].

6.1.2.2 Light-emitting polymer space-charge-limited transistor

To verify the concept of a light-emitting polymer SCLT, a top emitting PLED in the structure of Al/MoO_3(3nm)/PEDOT:PSS/Super Yellow/Ba(10 nm)/Ag(15 nm) was first prepared. Figure 6.15 shows the characteristics of the top emitting PLED. Super Yellow by Merck is a poly(para-phenylene vinylene) polymer derivative. The Al/MoO_3/PEDOT:PSS served as anode while Ba/Ag served as a transparent cathode. The maximum current efficiency was 10.6 cd/A and the luminescence exceeded 1000 cd/m^2 with 6 V. The characteristics of the top emitting PLED are suitable for integration with a polymer SCLT.

In the light-emitting transistor, the PLED was fabricated atop the Al collector of SCLT [11] as shown in Fig. 6.16. The energy band profile of the top-emitting polymer light-emitting transistor is shown in Fig. 6.17. Light emission and driving were separated into different components of the device with the collector (or anode) assigned a floating voltage between them. The floating voltage of the collector was determined by the voltage drop across the PLED. The performance of each component was, therefore, not compromised by the integration with other components. This structure was effectively a PLED connected in series to a variable resistor, modulating the PLED current and thus the luminance.

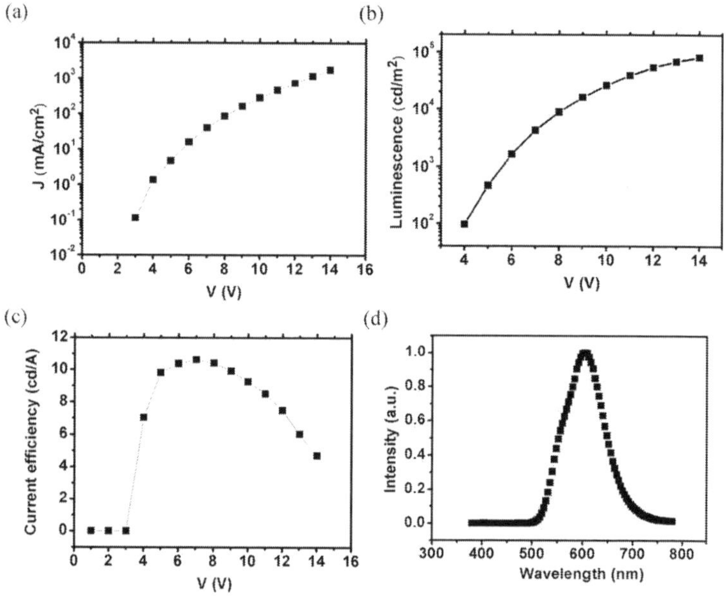

Figure 6.15 (a) Current density; (b) luminance; (c) current efficiency; and (d) spectrum of the top-emitting PLED [11].

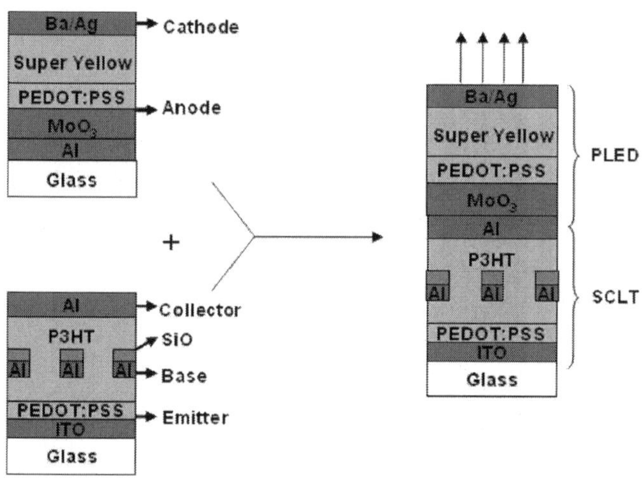

Figure 6.16 The structure of a light-emitting space-charge-limited transistor. A polymer light-emitting diode is stacked on a space-charge-limited transistor to form a light-emitting space-charge-limited transistor.

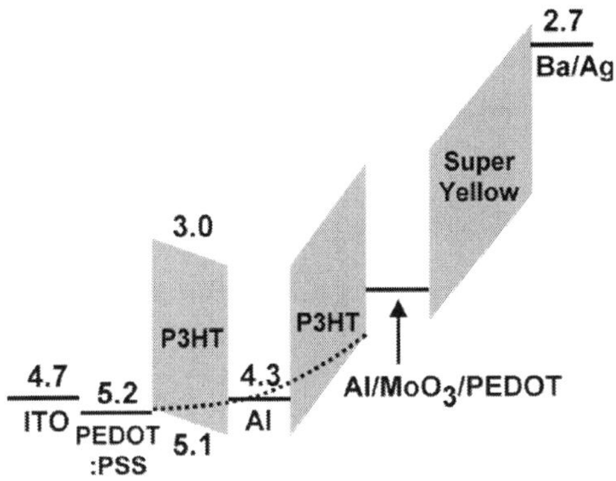

Figure 6.17 Energy profile of the top-emitting polymer light-emitting space-charge-limited transistor [11].

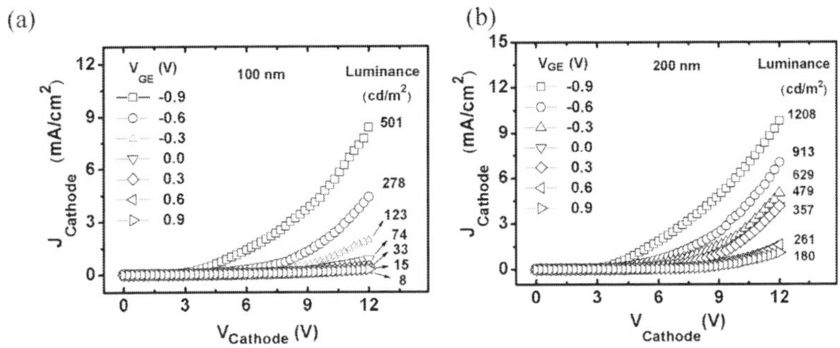

Figure 6.18 Characteristics of polymer light-emitting space-charge-limited transistor with (a) grid-opening diameter of 100 nm; and (b) 200 nm. The luminance output from the light-emitting transistor as V_{Cathode} based on −12 V is also indicated [11].

The voltage applied to the semitransparent cathode V_{Cathode} varied between 0 and −12 V while keeping emitter grounded. Figure 6.18 shows the characteristics of the transistor of the polymer light-emitting transistor with grid openings of 100 nm and 200 nm in diameter [11]. The luminance at various V_G for V_{Cathode} at −12 V is also indicated in Fig. 6.18. The maximum luminance achieved using SCLT with grid openings 100 nm and 200 nm in diameters was approximately 500 and 1200 cd/m^2. The luminance could be turned off by increasing the V_G. This device fulfilled the minimum luminance requirements for a display panel. The aperture ratio was about 100% because the light emitted upward was not shielded by the vertical metal-base transistor beneath.

6.2 Pressure Sensors

Artificial skin enables prosthetic limbs and robots to sense physical touch, temperature, or any other stimuli [14]. It can be stretched into various shapes, or be used to cover nonplanar surfaces when flexible sheets are used as a supporting substrate. Artificial skin is an integration of a large number of devices serving as an interface to convert outside information into electronic signals. Each pixel in an array requires a transistor for switching and driving, and sensors for monitoring information such as pressure [15, 16], temperature [17], and presence of particular molecules. In general, sensors can be viewed as variable resistors tuned by external information, as shown in Fig. 6.19. The information is read by the current passing through the device, while the transistor is turned on.

6.2.1 *Pressure Sensor Based on Organic Field-Effect Transistors*

Organic transistors have many advantages in sensor applications. They can be manufactured on plastic films at room temperature. They are mechanically flexible and potentially inexpensive to

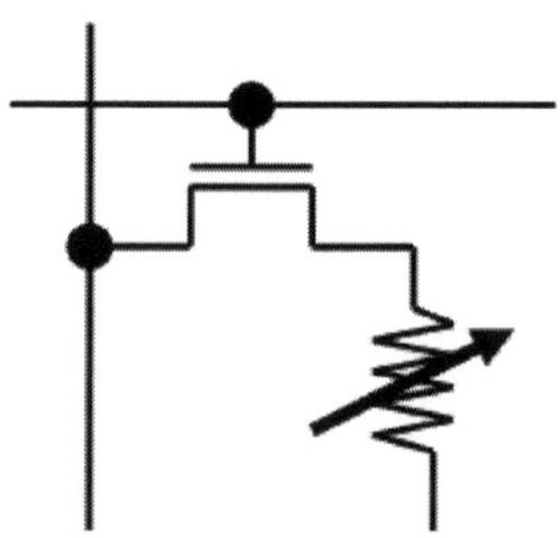

Figure 6.19 One pixel integrated device circuit.

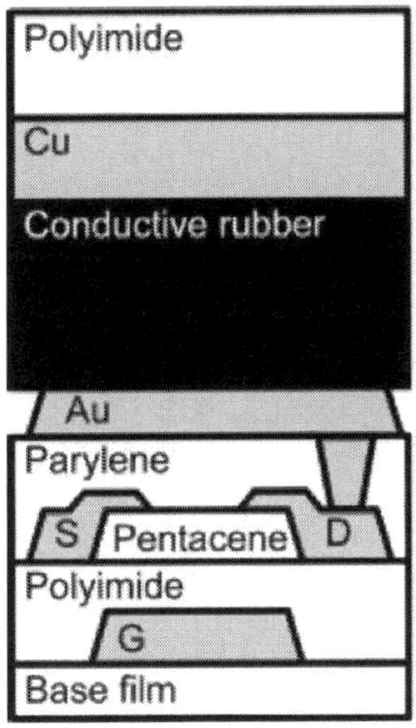

Figure 6.20 Cross-section of device structure of the pressure sensor integrating pressure sensing rubber with organic field-effect transistors [16].

manufacture. The combination of pressure sensing rubber and organic field-effect transistor active matrices has been demonstrated in Fig. 6.20 [16]. Cut-and-past customization of pressure sensor arrays has also been accomplished [18]. However, a via hole was required to make an electrical connection between the pressure sensing layer and the drain electrode of the transistor. Mechanical drills or laser drilling machines were commonly used to fabricate via holes; however, both processes add considerably to the complexity of the process. In addition, due to the low mobility of the organic materials employed and the long channel length between the source and drain electrodes, output current tends to be low and operational voltage high. Creating a transistor that is free from these drawbacks, is essential for the development of artificial skin. In the last section, polymer SCLT have been developed in order to realize a vertical transistor with low operational voltage. The integration of SCLT with pressure sensitive rubber was relatively easy, circumventing the need for a via hole, because both structures can be produced as a sandwich. This makes SCLT suitable for integration with pressure sensitive rubber to serve as a pixel within a pressure sensor operating at low voltage.

6.2.2 *Pressure Sensor Based on Polymer Space-Charge-Limited Transistors*

A pressure sensor has been realized by integrating a SCLT with a pressure sensitive rubber [19]. Figure 6.21 shows the structure of a pressure sensor comprising pressure sensitive rubber and a SCLT [19]. The pressure sensitive rubber, functioning as a variable resistor, is a composite of poly(dimethylsiloxane) (PDMS) and graphite particles. When no pressure is applied to the pressure sensitive rubber, the graphite particles provide conductive paths through the rubber as shown in Fig. 6.21a. The pressure sensitive rubber is in a low resistance state. When pressure is applied to the rubber, the PDMS deforms and the distribution of the graphite particles with the PDMS changes, as shown in Fig. 6.21b. The application of external pressure forces the graphite particles out of alignment with the conductive paths, changing the pressure sensitive rubber to a high resistance state. Voltage V_C, applied on the ITO atop the pressure sensitive rubber causes a floating $V_{C'}$ on the collector of the SCLT and the magnitude of $V_{C'}$ depends on the

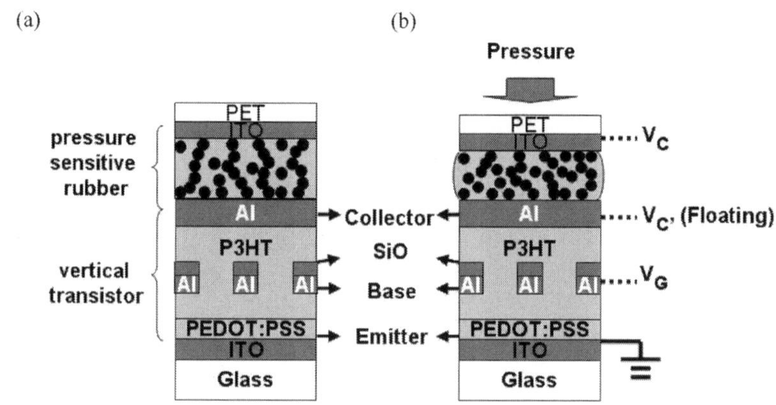

Figure 6.21 Device structure of the pressure sensor. Pressure sensitive rubber is vertically integrated with SCLT (a) without; and (b) with applied pressure [19].

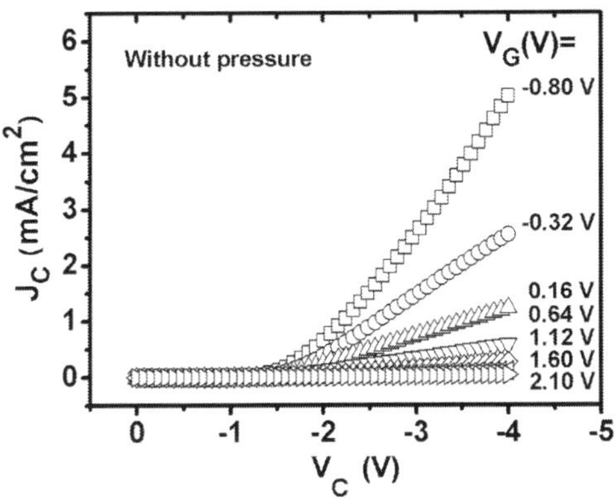

Figure 6.22 Output current from the integrated pressure sensor without applied pressure [19].

resistance of the pressure sensitive rubber. When the SCLT is turned on, the output current from the integrated pressure sensor reflects the magnitude of the pressure applied.

Figure 6.22 shows the characteristics of the pressure sensor without applying pressure [19]. Apparent modulation of output current was observed. When the transistor was in the on state, the output current density was 5.02 mA/cm^2. When V_C was -3.12 V, the on/off ratio of 132 was obtained.

Following the application of pressure on the integrated pressure sensor, a proportion of the conductive paths were broken, resulting in a change in resistance value, and hence, a change in output current [19]. Figure 6.23 shows the response of J_C under various pressure values. When the SCLT was turned on and pressures of 2.84 and 4.27 psi were applied, J_C was decreased, but remained at the same value until the pressure was changed. When the pressure was decreased, J_C recovered to close to its original value.

The reproducibility, stability, and response time of the pressure sensors were also investigated, to demonstrate the practical applicability of this device concept. First, the reproducibility of the pressure sensor was demonstrated according to the electrical characteristics of another two devices fabricated using the same procedures. Figure 6.24 shows the electric characteristics of the other two devices without applying pressure. These characteristics

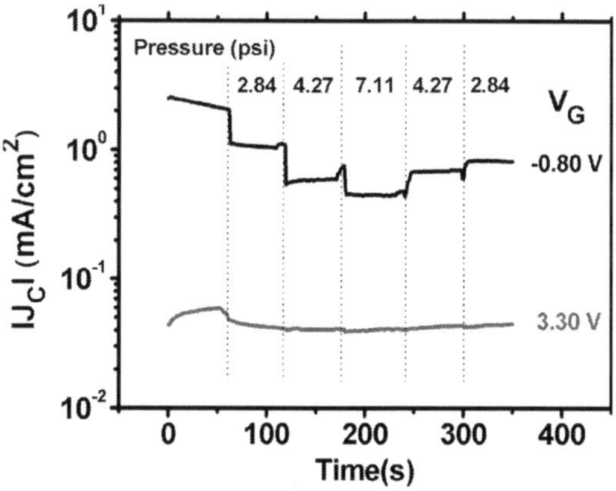

Figure 6.23 Response of J_C under various pressures. V_G was applied at −0.8 V and 3.3 V to switch on and off the SCLT. V_C was biased at −4 V [19].

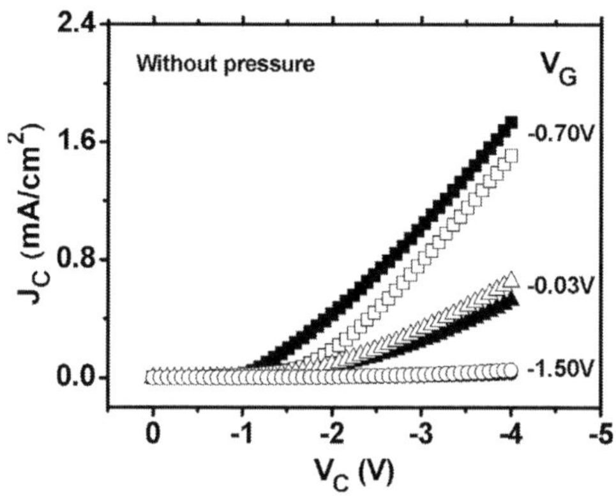

Figure 6.24 Electric characteristics of two additional pressure sensors without applied pressure. Solid and hollow symbols represent the data of different devices.

are similar to those shown in Fig. 6.22. When pressure was applied to these devices, the response of J_C was similar to that shown in Fig. 6.23. These results demonstrate the good reproducibility of the pressure sensor. Under certain amounts of pressure, the difference in the values of J_C between different devices was within 10%. These variations in J_C may have been the result of the characteristics of the pressure sensitive rubber. Second, the long-term stability of the

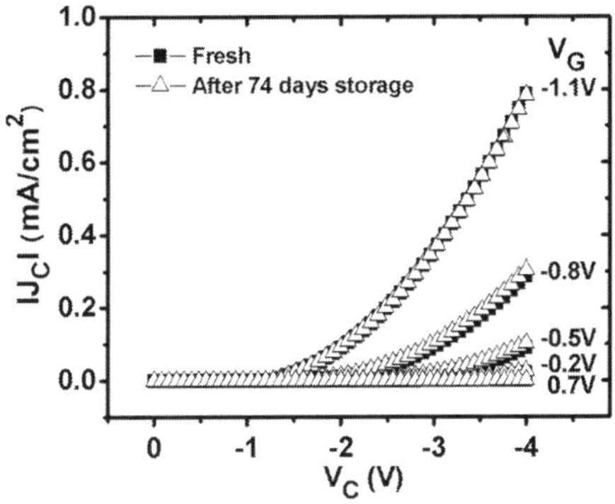

Figure 6.25 Characteristics of fresh device and device stored for 74 days.

pressure sensor was investigated by monitoring the characteristics of the transistors over time. Another device was prepared to study long-term stability, in which the characteristics of a fresh transistor were compared with those after being stored for 74 days, as shown in Fig. 6.25. Only slight differences were observed.

Finally, the response time of the pressure sensitive rubber was investigated by recording the current response using the circuit shown in Fig. 6.26a. The pressure sensitive rubber was connected in series with a 10 k ohm resistor to identify variations in the resistance of the pressure sensitive rubber. As shown in Fig. 6.26b,

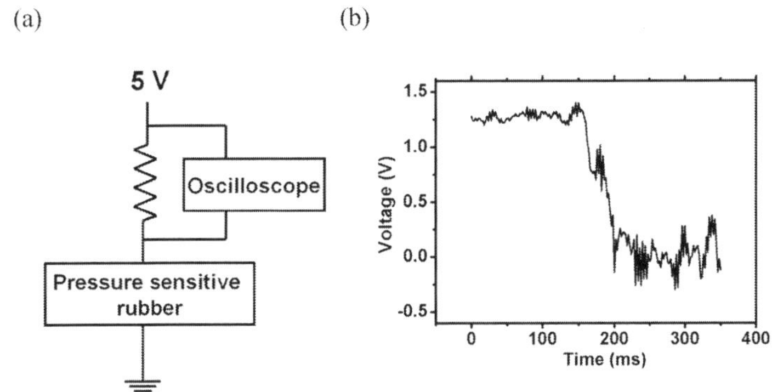

Figure 6.26 (a) Circuit for measuring current response; (b) current response of pressure sensitive rubber [19].

when touching the pressure sensitive rubber carefully by hand, the voltage dropped to a value of nearly zero within 50 ms. It required approximately 22 ms for the voltage across the resistor to drop to 0.625 V.

6.3 Proximity Sensors

Developing robots capable of interacting with humans and moving in unstructured environments poses a number of fascinating challenges. In fast changing environments, robots would be capable of preventing collisions by sensing their environment in real-time. To accomplish this goal, the surface of robots could be covered with an array of proximity sensors, enabling the robot to move in unstructured and unpredictable environments, such as homes, crowded streets, or hospitals, without the threat of collision [20].

Conventional proximity sensors are based on inorganic near infrared light-emitting diodes and photo-detectors. Without the need of physical contact, proximity sensors are able to detect the presence of nearby objects. Unfortunately, these devices cannot be integrated through monolithic fabrication as an array on a flexible substrate, which would be an ideal solution for the surface of robots or other machines. Electronic devices based on a flexible substrate, such as organic electronic devices, could be used for the development of sensitive skin to cover moving machines like robots or cars. Unlike conventional approaches of inorganic devices, fabricating polymer light-emitting diodes and polymer photo-detectors on a plastic substrate with high pixel density would be relatively easy. Organic optoelectronic devices based on conjugated polymers could be prepared using a solution process on a large-area flexible substrate to meet the requirements of sensitive skin.

Recently, a polymer proximity sensor operating in the visible range was reported [21]; however, the detection distance was below about 15 mm, which is far below the requirements for robot applications. Since light scattering occurs proportionally to the inverse of the fourth power of its wavelength, light in the infrared range does not suffer from scattering seriously. Therefore, the light in the infrared range would be far preferable for proximity sensors.

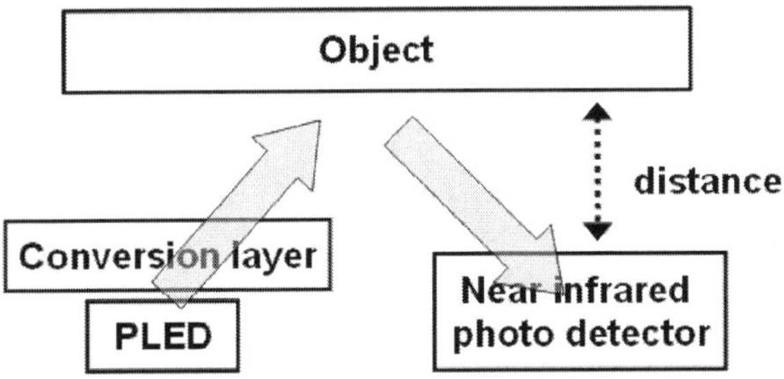

Figure 6.27 Working principle of near infrared proximity sensor.

The excitation light and the reflected light in the infrared range are much stronger than that of visible light. A polymer infrared proximity sensor has been produced using a near infrared light source and a near infrared photo detector [22, 23]. The device structure of the polymer infrared proximity sensor is shown in Fig. 6.27. The near infrared light source was a compact device integrating a polymer light-emitting diode with a color conversion film composed of a polymer host and an infrared laser dye. An inorganic phosphor could also be used to replace laser dye. The near infrared photo detector was a polymer donor-acceptor blend, sensitive to near infrared photons [24, 25]. In the following sections, a description of each component of the compact proximity sensor is presented in detail.

6.3.1 *The Near Infrared Polymer Photo Detector*

A blend of P3HT and (6,6)-phenyl-C61-butyric acid methyl ester (PCBM) is conventionally used as the active material in a polymer solar cell or polymer photo detector. As shown in Fig. 6.28, the photo detector with the thickness less than 4 μm only shows the sensitivity to visible light [24]. When the thickness of the active material was increased, the photo response shifted to the infrared range. The primary mechanism associated with the change in photoresponse was described by addressing the visible and infrared parts individually [24]. For the absorption in the visible range, the absorption was through the P3HT, in which most photons were absorbed within the thickness of 200 nm, resulting in high

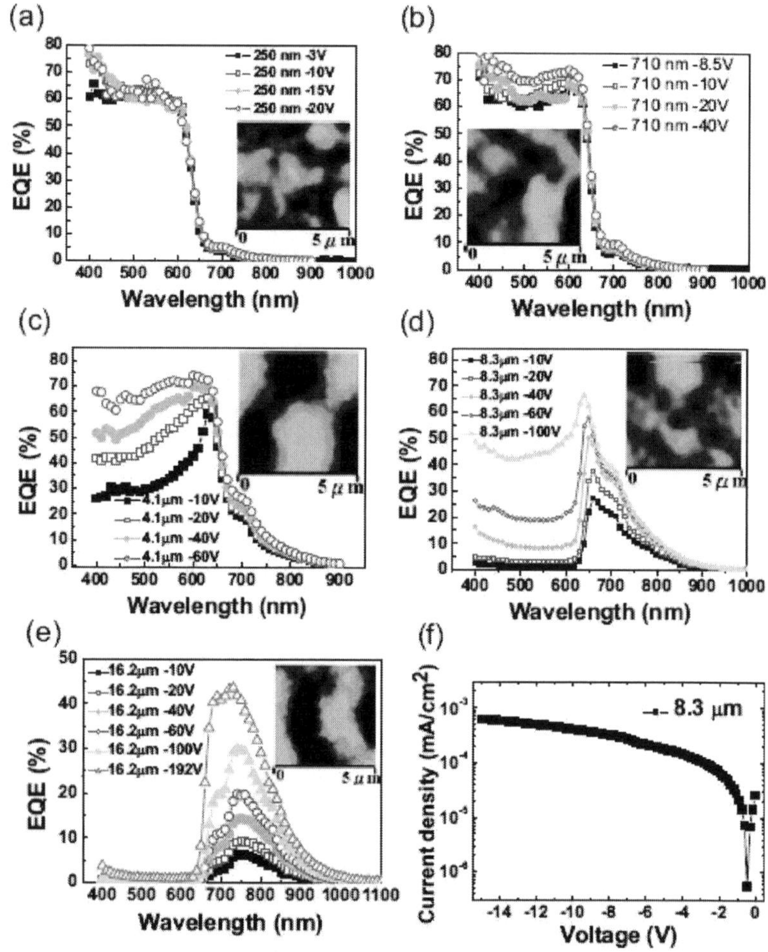

Figure 6.28 External quantum efficiency and atomic force microscopic images of photo detector with active materials of various thicknesses [24].

external quantum efficiency values of the device with 250 and 710 nm active materials. When the thickness of the active materials were increased, it was difficult for the carriers dissociated from the excitons near the anode side to move across the bulk and be collected by the cathode to generate a photocurrent. Most carriers recombined with each other in the bulk, thereby reducing the external quantum efficiency in the visible range, following an

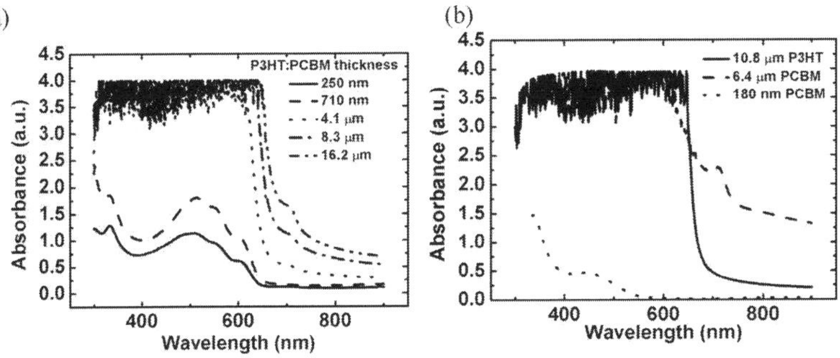

Figure 6.29 The absorption of (a) P3HT:PCBM of various thickness; and (b) net film of P3HT and PCBM [24].

increase in thickness. As for the absorption in the infrared area, the photons in the infrared range was described to be absorbed by the thick PCBM, thereby increasing the external quantum efficiency, following an increase in the thickness of the active layer [24]. The absorption spectra of the P3HT, PCBM, and P3HT:PCBM films of various thicknesses are shown in Fig. 6.29, in which most visible light is absorbed by all of the devices with different thicknesses, and differences in absorption could be observed in the near infrared range (650–1000 nm). The absorption intensity in the infrared range increased with an increase in film thickness. The absorption spectra of thick pure P3HT and PCBM layer were measured and shown in Fig. 6.29 in order to study the origin of the infrared absorption. The absorption of PCBM layer in the infrared range was observed to be increased dramatically following an increase in the thickness of PCBM to 6.4 μm, while no absorption by P3HT layer was observed in the infrared range, even with a film thickness of 10.8 μm. The P3HT in the active layer was used only to create a heterojunction for exciton dissociation and the transport of charge. Therefore, a device with a thick P3HT:PCBM active layer was demonstrated to be suitable as a near infrared photo detector.

6.3.2 *The Near Infrared Polymer Light Source*

It was demonstrated that a near infrared light can be generated by converting the light of a polymer light-emitting diode in the visible

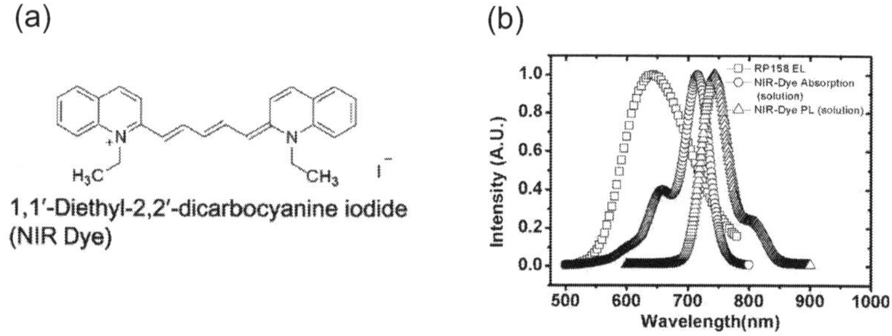

Figure 6.30 (a) Molecular structure of the near infrared dye; (b) Electroluminescence of RP158 and absorption and photoluminescence of near infrared dye [22].

range to the infrared range [22]. This can be achieved by using a color conversion film consisting of near infrared dye and a host polymer [22].

The molecular structure of near infrared dye is shown in Fig. 6.30a, while the photoluminescence (PL) and absorption spectra of the near infrared dye in a solution and in a film as well as the EL spectrum of the red polymer light-emitting diode are shown in Fig. 6.30b. RP158 is used as the active material in red polymer light-emitting diodes. Obviously, there is a major overlap exists between the EL of the red polymer and the absorption of the near infrared dye, resulting in an efficient near infrared emission, with the peak of the near infrared emission at 750 nm. Since the dye concentration is low in the conversion film, the red emission was approximately the same as the near infrared emission even though the color conversion film was as thick as 10 μm. Color conversion film could also be prepared using inorganic phosphor R670 [23]. Figure 6.31a shows the absorption and PL spectrum of the of R670 phosphor powder. The broad absorption of inorganic phosphor R670 makes it with a wide excitation range between 200 and 640 nm. The emission of inorganic phosphor R670 peaked at 670 nm extending up to 800 nm. The PL and absorption spectra of the R670 conversion layer as well as the EL spectrum of the green polymer light-emitting diode are shown in Fig. 6.31b. There is also an overlap exists between the EL of the polymer light-emitting diode and the absorption of the phosphor, resulting in an efficient near infrared emission.

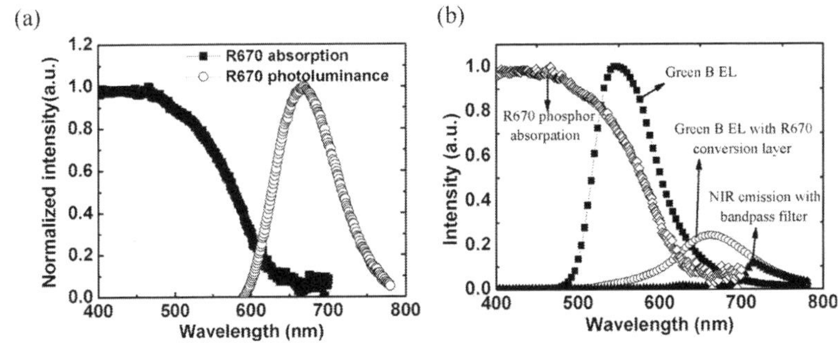

Figure 6.31 (a) Absorption and photoluminescence of phosphor R670 in the conversion layer; (b) Electroluminescence spectrum of the polymer light-emitting diode made using Green B converted to the near infrared range using an R670 conversion layer [23].

With the help of a band-pass filter, the operational wavelength was maintained between 700 and 800 nm, with a peak at 716 nm.

6.3.3 *The Polymer Proximity Sensor*

As shown in Fig. 6.27, a red RP158 polymer light-emitting diode, near infrared dye conversion layer, and polymer photo detector were first placed side-by-side to demonstrate the idea of the polymer proximity sensor [22]. The response is shown in Fig. 6.32. The maximum detection distance was 19 cm for white and red papers,

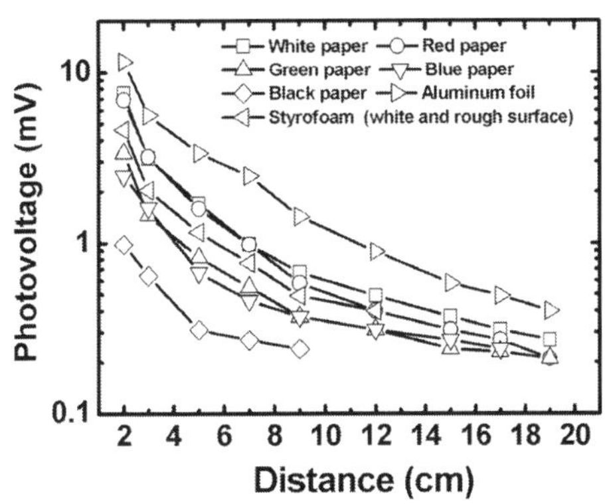

Figure 6.32 Response of the detection as a function of distance [22].

Figure 6.33 Proximity sensor array combining 9 photodetectors with 16 PLEDs on the circuit board [23].

17 cm for green and blue, and 9 cm for black. It was reported that the maximum detection distances were 15 cm for Styrofoam and 19 cm for aluminum foil. Although the detection distance was affected by the color and surface texture of the object, the detection distances for all materials was greater than 9 cm. When using a green polymer light-emitting diode, R670 phosphor conversion layer, and a polymer photo detector, similar results were also demonstrated [23]. Figure 6.33 shows an array integrated with 9 photo detectors and 16 polymer light-emitting diodes. Polymer light-emitting diodes, polymer photodetectors and the amplifiers were integrated on a single printed circuit board, as shown in Fig. 6.33. In order to demonstrate the objects can be detected instantly, the software was designated to show nine blocks on the screen representing the nine photo detectors on the circuit board. The block became white as objects approached; otherwise remaining black. The real-time detection of a 4 cm by 4 cm white paper moving from one side to another continuously has been demonstrated as shown in Fig. 6.34. It was also successful when using other moving objects to demonstrate the real-time detection. The polymer proximity sensor array detected the motion of objects in a real-time manner.

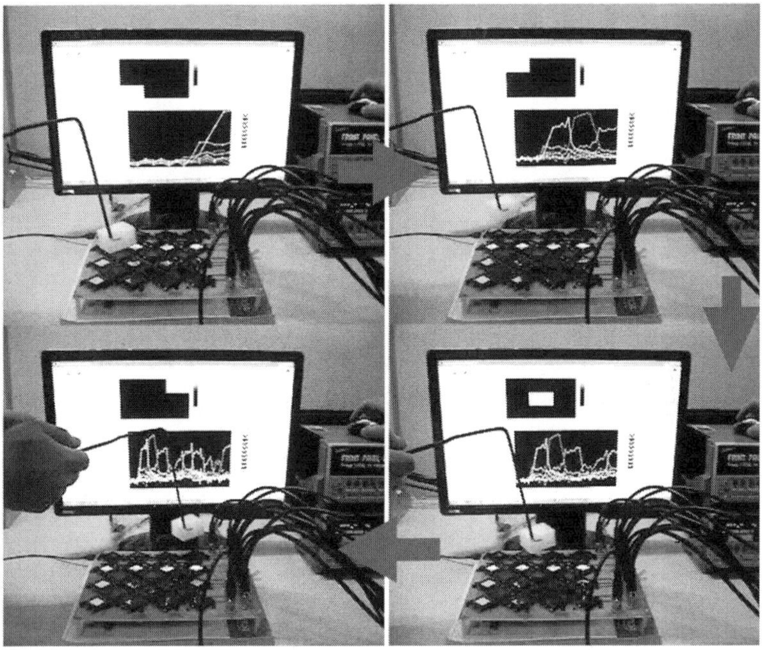

Figure 6.34 As the object is moved above the array, its position is instantly detected and displayed on the computer screen as the white block. Each bock corresponds to one photo detector in the array [23].

The information was fed back to the processing unit of the robot to respond.

References

1. L. Zhou, S. Park, B. Bai, J. Sun, S. C. Wu, T. N. Jackson, S. Nelson, D. Freeman, and Y. Hong (2005) *IEEE Electron Dev. Lett.* **26**, 640.
2. J. H. Cheon, J. H. Choi, J. H. Hur, J. Jang, H. S. Shin, J. K. Jeong, Y. G. Mo, and H. K. Chung (2006) *IEEE Trans. Electr. Dev.* **53**, 1273.
3. C. L. Lin and Y. C. Chen (2007) *IEEE Electron Dev. Lett.* **28**, 129.
4. S. Ono, K. Miwa, Y. Maekawa, and T. Tsujimura (2007) *IEEE Trans. Electr. Dev.* **54**, 462.
5. M. L. Chabinyc, F. Endicott, B. D. Vogt, D. M. DeLongchamp, E. K. Lin, Y. Wu, P. Liu, and B. S. Ong (2006) *Appl. Phys. Lett.* **88**, 113514.

6. C. J. Yang, C. L. Lin, C. C. Wu, Y. H. Yeh, C. C. Cheng, Y. H. Kuo, and T. H. Chen (2005) *Appl. Phys. Lett.* **87**, 143507.
7. F. Cicoira and C. Santato (2007) *Adv. Funct. Mater.* **17**, 3421.
8. J. Zaumseil and H. Sirringhaus (2007) *Chem. Rev.* **107**, 1296.
9. J. Cornil, J.-L. Brédas, J. Zaumseil, and H. Sirringhaus (2007) *Adv. Mater.* **19**, 1791.
10. J. Zaumseil, C. L. Donley, J. S. Kim, R. H. Friend, and H. Sirringhaus (2006) *Adv. Mater.* **18**, 2708.
11. C. Y. Chen, Y. C. Chao, H. F. Meng, and S. F. Horng (2008) *Appl. Phys. Lett.* **93**, 223301.
12. Y. C. Chao, H. F. Meng, and S. F. Horng (2006) *Appl. Phys. Lett.* **88**, 223510.
13. Y. C. Chao, H. F. Meng, S. F. Horng, and C. S. Hsu (2008) *Org. Electron.* **9**, 310 (2008).
14. T. Sekitani and T. Someya (2010) *Adv. Mater.* **22**, 2228.
15. T. Someya, T. Sekitani, S. Iba, Y. Kato, H. Kawaguchi, and T. Sakurai (2004) *Proc. Nat. Acad. Sci.* **101**, 9967.
16. T. Someya, Y. Kato, T. Sekitani, S. Iba, Y. Noguchi, Y. Murase, H. Kawaguchi, and T. Sakurai (2005) *Proc. Nat. Acad. Sci.* **102**, 12321.
17. S. Jung, T. Ji, and V. K. Varadan (2006) *Appl. Phys. Lett.* **90**, 062105.
18. H. Kawaguchi, T. Someya, T. Sekitani, and T. Sakurai (2005) *IEEE J. Solid-State Circuits* **40**, 177.
19. Y. C. Chao, W. J. Lai, C. Y. Chen, H. F. Meng, H. W. Zan, and S. F. Horng (2009) *Appl. Phys. Lett.* **95**, 253306.
20. V. J. Lumelsky, M. S. Shur, and S. Wagner (2001) *IEEE Sens. J.* **1**, 41.
21. L. Bürgi, R. Pfeiffer, M. Mücklich, P. Metzler, M. Kiy, and C. Winnewisser (2006) *Org. Electron.* **7**, 114.
22. E. C. Chen, S. R. Tseng, J. H. Ju, C. M. Yang, H. F. Meng, S. F. Horng, and C. F. Shu (2008) *Appl. Phys. Lett.* **93**, 063304.
23. E. C. Chen, C. Y. Shih, M. Z. Dai, H. C. Yeh, Y. C. Chao, H. F. Meng, H. W. Zan, W. R. Liu, Y. C. Chiu, Y. T. Yeh, J. C. Liao, C. J. Sun, S. F. Horng, and C. S. Hsu (2011) *IEEE T. Electron Dev.* **58**, 1215.
24. E. C. Chen, C. Y. Chang, J. T. Shieh, S. R. Tseng, H. F. Meng, C. S. Hsu, and S. F. Horng (2010) *Appl. Phys. Lett.* **96**, 043507.
25. C. M. Yang, P. Y. Tsai, S. F. Horng, K. C. Lee, S. R. Tzeng, H. F. Meng, J. T. Shy, and C F. Shu (2008) *Appl. Phys. Lett.* **92**, 083504.

Index

Color Insert

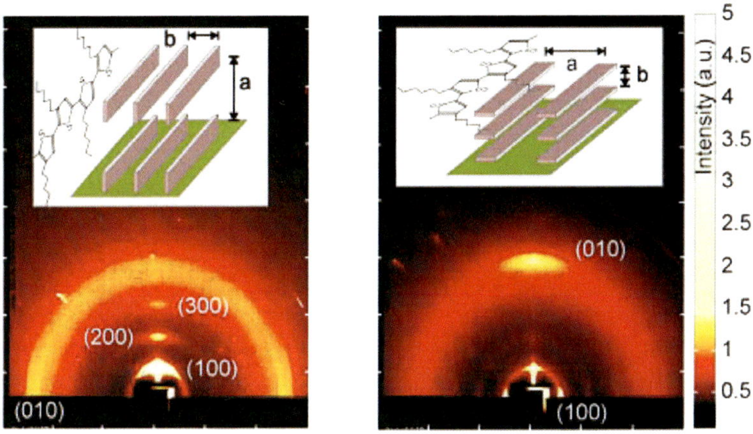

Figure 1.20

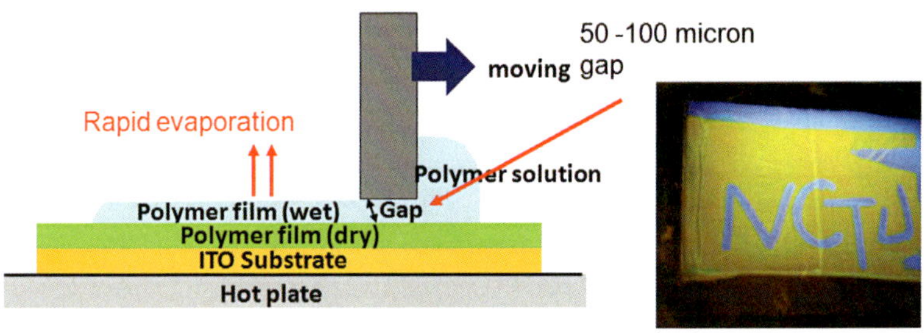

Figure 2.5

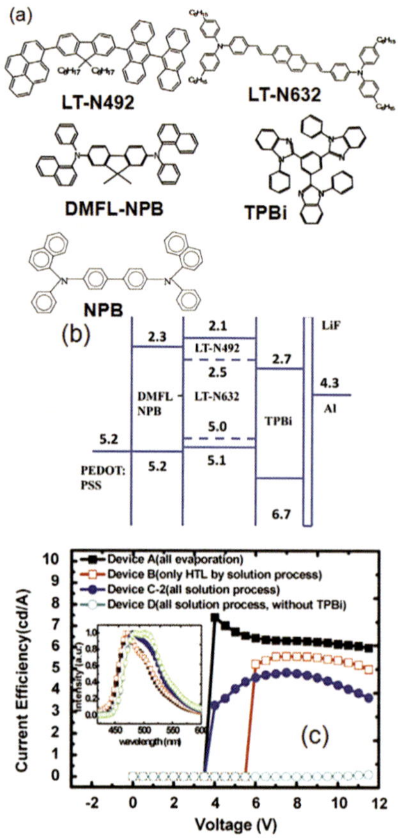

Figure 2.6

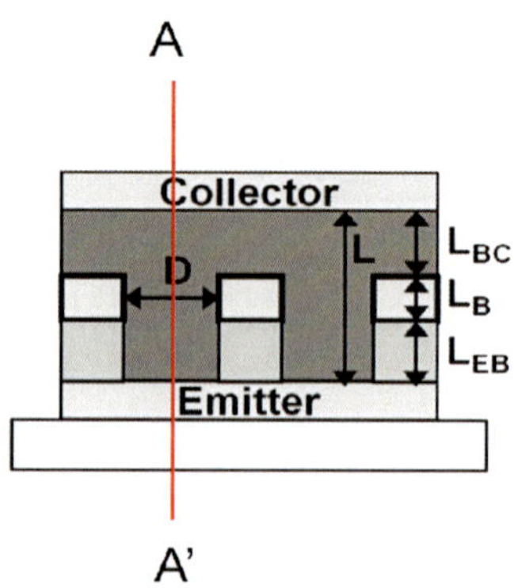

Figure 4.17

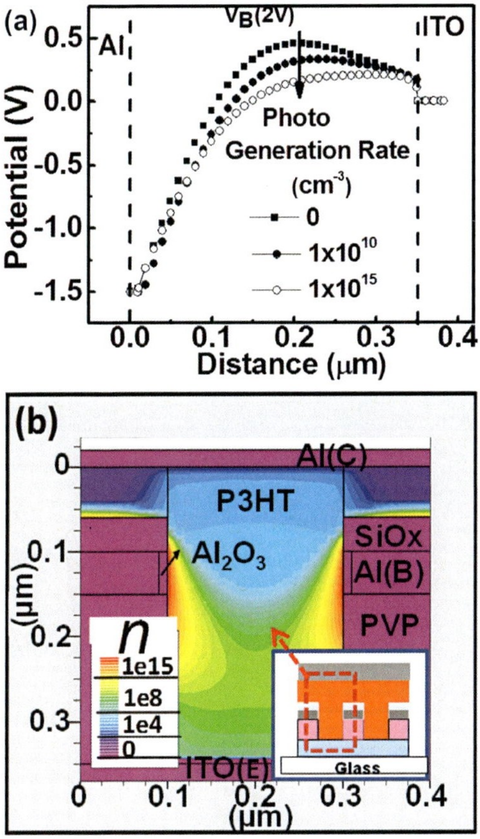

Figure 4.57

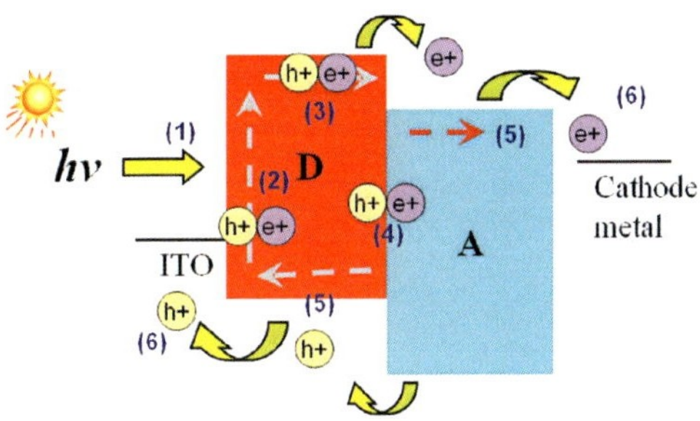

Figure 5.1

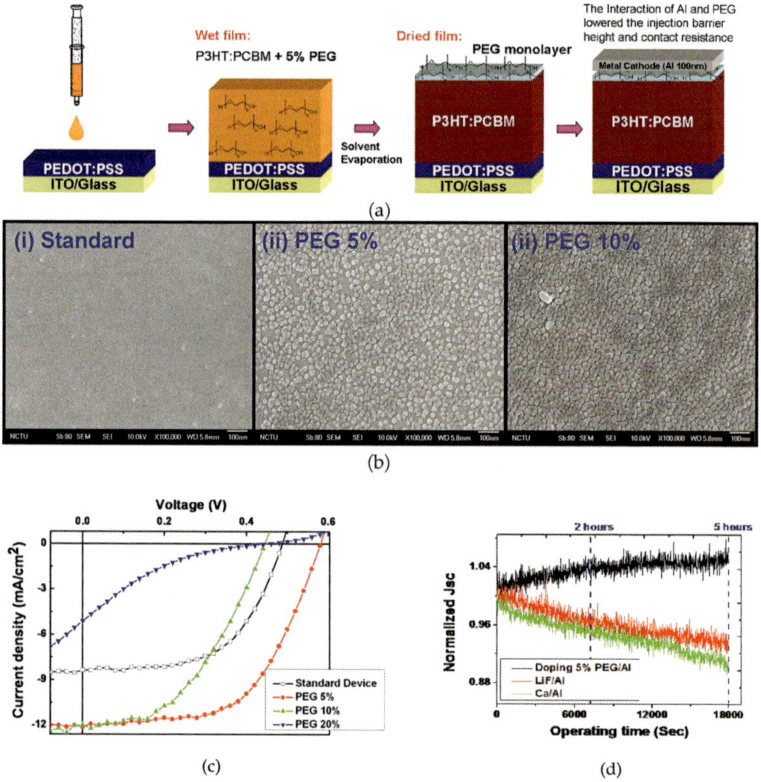
Wet film:
P3HT:PCBM + 5% PEG
Dried film:
PEG monolayer
The Interaction of Al and PEG lowered the injection barrier height and contact resistance
Metal Cathode (Al 100nm)
Solvent Evaporation
P3HT:PCBM
PEDOT:PSS
ITO/Glass
(a)
(i) Standard
(ii) PEG 5%
(ii) PEG 10%
(b)
Voltage (V)
Current density (mA/cm2)
Standard Device
PEG 5%
PEG 10%
PEG 20%
(c)
2 hours
5 hours
Normalized Jsc
Operating time (Sec)
Doping 5% PEG/Al
LiF/Al
Ca/Al
(d)

Figure 5.10